T0137208

AAPS Advances in the Pharmaceutical Sciences Series

Volume 40

Series Editor

Yvonne Perrie, Strathclyde Institute of Pharmacy and Biomedical Sciences, University of Strathclyde, Glasgow, UK

The AAPS Advances in the Pharmaceutical Sciences Series, published in partnership with the American Association of Pharmaceutical Scientists, is designed to deliver volumes authored by opinion leaders and authorities from around the globe, addressing innovations in drug research and development, and best practice for scientists and industry professionals in the pharma and biotech industries.

More information about this series at http://www.springer.com/series/8825

James E. De Muth

Practical Statistics for Pharmaceutical Analysis

With Minitab Applications

James E. De Muth
School of Pharmacy
University of Wisconsin-Madison
Madison, WI, USA

ISSN 2210-7371   ISSN 2210-738X (electronic)
AAPS Advances in the Pharmaceutical Sciences Series
ISBN 978-3-030-33991-3   ISBN 978-3-030-33989-0 (eBook)
https://doi.org/10.1007/978-3-030-33989-0

This Springer imprint is published by the registered company Springer Nature Switzerland AG
The registered company address is: Gewerbestrasse 11, 6330 Cham, Switzerland

Dedicated to
Grayson and Korra

Preface

The purpose of this book is to present a brief and organized approach to statistics for scientists working in the pharmaceutical industry, as well as an introduction to statistics for undergraduate and graduate students in the pharmaceutical sciences. It is designed for individuals desiring a brief introduction to the field of statistics, as well as a quick reference for statistical problem-solving. It can serve as a guide and a reference for researchers in need of methods to statistically analyze data.

Unlike other texts, which are heavy with mathematical formulas, this book emphasizes the selection of the correct statistical test to address specific types of data and the interpretation of those statistical results. The primary objective is to help researchers identify the most appropriate test given the type of variables present in a study. To assist with this process, Appendix A provides a flow chart to help determine the best test for any given situation.

A Book for Non-statisticians

Statistics provide useful methods to analyze data commonly seen in the laboratory. Evaluating these findings helps in making correct and beneficial decisions. These tests provide a means for answering questions faced by analytical scientists working in the pharmaceutical industry. With an organized approach to evaluating observed data, it helps researchers avoid jumping to conclusions and making choices that may be unwise or even dangerous to the quality of the product or application being tested.

During the 2004 Land O'Lakes Bioanalytical Conference at the University of Wisconsin, Wendell Smith spoke on the topic of statistical issues with developing and validating methods. He made two interesting observations: (1) statistics "provides methods and tools for decision making in the face of uncertainty," and (2) successful applications of statistics "requires collaboration among participating scientists." Regarding the first statement, data will vary (the uncertainty), and statistics are useful tools that can help control and interpret this variability. This book deals with that uncertainty. Possibly of greater importance is the second statement. By participating scientists, Smith was referring to the "analytical" scientist and the "statistical" scientists, both of whom are required to accomplish a successful statistical analysis. Some questions and decisions are clearly in the analytical domain and need to be addressed by the scientist designing the study and collecting the data. Other questions and applications need to be the focus for professional statisticians. Working together in collaboration, a successful evaluation of sample data can be accomplished using statistics. At certain points in this book, the appropriate delegation of these responsibilities will be noted.

Contents of This Book

The book is still divided into two major sections: (1) the underpinnings required to understand the basic elements of statistical tests (terminology, descriptive statistics, statistical inference, and hypothesis testing) and (2) inferential statistics to help in problem-solving (test identifying differences, relationships, similarities, and potential outliers). Additional supportive materials are provided in the Appendices with flow charts, tables, software commands, and proofs for statistical calculations.

Chapter 1 focuses on the terminology and essential elements of statistical testing. Statistics is often complicated by synonyms, and this chapter establishes the terms used in the book and how rudiments interact to create statistical tests. Chapter 2 presents descriptive statistics that are used to organize and summarize sample results. Chapter 3 introduces the basic concepts of probability, the characteristics of a normal distribution, the alternative approaches for non-normal distributions, and the topic of making inferences about a larger population based on a small sample taken from that population. Chapter 4 discusses hypothesis testing where computer output is interpreted and decisions are made regarding statistical significance. This chapter also deals with the determination of appropriate sample sizes.

The next four chapters focus on tests that make decisions about a population based on a small subset of that population. Chapter 5 looks at statistical tests that evaluate whether a significant difference exists in a population based on the results of sample data. In Chap. 6, the tests try to determine the extent and importance of relationships between variables being compared. In contrast to Chap. 5, Chap. 7 presents tests that evaluate the equivalence or similarity among samples, instead of the difference between the samples (how close samples are to being the same). The last chapter deals with potential outlier or aberrant values and how to statistically determine if they should be removed from the sample data.

Each statistical test presented includes at least one example problem with the resultant software output and how to interpret the results. Minimal time will be spent on the mathematical calculations or theory. For those interested in the mechanics of the associated equations, supplemental figures will be presented for tests with their respective formulas. In addition, Appendix D presents the equations and mathematical proofs for the output result for many of the various examples.

Minitab Software

Examples and results from the appropriate statistical results will be displayed using Minitab[1]. In addition to the results, the required steps to set up and analyze data using Minitab will be presented with the examples for those having access to this software.

Numerous other software packages are available, including basic statistical programs available on Excel. Minitab was selected by the author because of the ease of use, extent of available statistical tests, and quality of output reports. It is a relatively simple point-and-click technology to help scientists with a variety of data analysis problems including design of experiment, system analysis, quality tools, graphic analysis, and process analytical control. This book only focuses on the Minitab applications for descriptive and inferential statistics.

Guidelines for the Use of This Book

The goals for this book are twofold: (1) to serve as a useful resource for selecting and interpreting inferential statistics and (2) to provide a practical guide for using Minitab software to evaluate sample data.

If the reader is interested in learning more about statistics and not interested in Minitab, simply read the text covering the importance of these tests, the required variables, and the critical assumptions. Look at the example problems, the computer output results (which will be similar to most commercial statistical packages), and how to interpret these results. Ignore the sections covering Minitab. However, if the reader has access to Minitab, procedures are described for the individual descriptive and inferential statistics.

As mentioned previously, this book is designed to minimize the time spent on mathematical formulas. The main text describes when each test should be used, the type of variables required, and any special rules or conditions for the use of each test. For those interested, the mathematical formulas are presented as figures, but their derivation is not discussed. Have faith, the equations work! Examples illustrate the results seen with the computer software evaluation of the data and how to interpret the results of the various tests. Proofs that the software and hand calculation are the same (worked out for most examples) are presented in Appendix D for those interested in the mathematical manipulations.

---

[1] Portions of information contained in this publication/book are printed with permission of Minitab, LLC. All such material remains the exclusive property and copyright of Minitab, LLC. All rights reserved.

MINITAB® and all other trademarks and logos for the Company's products and services are the exclusive property of Minitab, LLC. All other marks referenced remain the property of their respective owners. See minitab.com for more information.

Acknowledgments

The author has been fortunate to have taught over 200 statistics short courses throughout the United States and internationally. The input from the learners attending these classes and examples of data they shared has helped create examples used in this book. In addition, the author has had the opportunity to work closely with a variety of excellent statisticians especially while serving as Chair of the USP Biostatistics Expert Committee. Both these students and colleagues have helped contribute to the author's understanding of statistics and to the development of the contents of this book.

Thanks to Taylor & Francis Group for releasing the intellectual property rights to the previous statistics book *Basic Statistics and Pharmaceutical Statistical Applications*. Very little material is duplicated in this book, but for those sections that are duplicated, Taylor & Francis Group's support of this effort is greatly appreciated. Various references in this book refer to sections in the aforementioned text, and if required by the reader, it covers selected topics in much greater detail.

The completion of this text is directly attributable to the love and support of my family. A very special thanks to my wife, Judy, and to our daughters, Jenny and Betsy, for their continued encouragement and patience during my reported "retirement."

Madison, WI, USA   James E. De Muth
December 2019

Statistical Symbols

α (alpha)	Type I error, probability used in statistical tables (p)
β (beta)	Type II error, population slope
$1 - \alpha$	Level of significance, level of confidence
$1 - \beta$	Power
δ (delta)	Difference
μ (mu)	Population mean
$\mu_{\bar{X}}$	Mean of the sampling distribution of $\bar{X}$
ν (nu)	Degrees of freedom, also symbol df (unusually $n - 1$)
ρ (rho)	Spearman rank correlation coefficient, population correlation coefficient
σ (sigma)	Population standard deviation
σ^2	Population variance
$\sigma_{\bar{X}}$	Standard deviation of the sampling distribution of $\bar{X}$
χ^2 (chi)	Chi square coefficient
$\cap$	Intercept symbol (probability of two events occurring)
$\cup$	Conjoint symbol (probability of either of two events occurring)
a	y-intercept, intercept of a sample regression line
b	Sample slope, slope of a sample regression line
c or C	Number of columns in a contingency table, number of possible comparison with two levels
CV or %CV	Coefficient of variation
df	Degrees of freedom, also symbol ν (usually $n - 1$)
E	Event or expected frequency with chi square
$\bar{E}$	Not event, compliment of E
F	F-statistic, analysis of variance coefficient, test statistic
H	Kruskal-Wallis test statistic
H_0	Null hypothesis, hypothesis under test
H_1	Alternate hypothesis, research hypothesis
IQR	Interquartile range
LEL	Lower equivalence limit
LTL	Lower tolerance limit

Log	Logarithm to the base 10
MS_B	Mean square between
MS_E	Mean squared error, synonym MS_W
MS_R	Mean squared residual
MS_{Rx}	Mean squared treatment effect
MS_W	Mean square within, synonym MS_E
n	Number of values or data points in a sample
N	Number of values in a population, total number of observations
$n!$	Factorial (product of all whole numbers from 1 to n)
O	Observed frequency with chi square
p	Probability, level of significance, type I error
$p(E)$	Probability of event E
$p(x)$	Probability of outcome x
Q_1	25th percentile
Q_2	50th percentile, synonym median
Q_3	75th percentile
r	Correlation coefficient, Pearson's correlation
r or R	Number of rows in a contingency table
r^2	Coefficient of determination
R^2	Coefficient of determination (Minitab)
RSD or %RSD	Relative standard deviation
S or SD	Sample standard deviation
S_p^2	Pooled variance
SE or SEM	Standard error of the mean, standard error
t	t-test statistic
U	Mann-Whitney U test statistic
UEL	Upper equivalence limit
UTL	Upper tolerance limit
x_i	Any data point or value
$\overline{X}$	Sample mean
$\overline{X}_G$	Geometric mean, grand mean
Z	Z-test statistic

Contents

Chapter 1
Essential Elements and Statistical Terms

Abstract The first chapter addresses some of the basic elements associated with statistical analysis and attempts to clarify some of the confusing language and synonyms. Statistics can be divided into two major types: (1) descriptive statistics to summarize sample data and (2) inferential statistics to make statements about an entire population from which the sample was taken. Presented will be steps to consider in performing a statistical procedure and important assumptions associated with all statistical tests. Identifying the types of variables in the study is critical because they will dictate the type of descriptive and/or inferential statistics that are most appropriate to test and report. Briefly discussed will be sampling plans, reportable values, rounding, and significant figures. Minitab is introduced as software to provide a rapid and efficient method for statistically analyzing sample data.

Keywords Descriptive statistics · Inferential statistics · Ordinal variable · Parameter · Population · Qualitative variable · Quantitative variable · Sampling · Sample

This chapter introduces many of the terms commonly seen in the literature, presented in scientific reports and available on statistical software. It focuses on population parameters versus sample statistics; descriptive versus inferential statistics; the importance of defining study variables; critical assumptions involved with inferential statistics; sample plans; and reportable values, significant figures, and rounding. One of the major problems with understanding statistics is the number of synonyms used by various authors. The chapter starts with the very basics and attempts to clarify some of this confusing language. The chapter ends with a discussion of the software used for the examples presented in the following chapters.

© American Association of Pharmaceutical Scientists 2019   1
J. E. De Muth, *Practical Statistics for Pharmaceutical Analysis*, AAPS
Advances in the Pharmaceutical Sciences Series 40,
https://doi.org/10.1007/978-3-030-33989-0_1

1.1   Descriptive Statistics vs. Inferential Statistics

Statistics can be divided into two major types. The first is *descriptive statistic*, where data collected by the analyst is organized and presented as a summary. For example, in Table 1.1 are the results of the percent label claim for 20 samples collected at random during the production run for an oral solid dosage form. As such, they show the results of the sample, but not much useful information. True, it is an accurate account for the samples but relatively meaningless other than 20 individual data points. Descriptive statistics can take this information and provide a measure of center for the distribution of these samples and an indication of the variability around this center. These types of descriptive statistics (e.g., mean and standard deviation) will be discussed in Chap. 2.

The second type of statistics are termed inferential statistics. The sample presented above represents only 20 observations out of a potential of thousands or millions of these oral dosage forms. Inferential statistics allow the researcher to predict the characteristics of the entire batch from which the sample is taken. This is illustrated in Fig. 1.1. Ideally we would know the entire population (batch) being produced, but this would be too expensive and impractical to measure every single unit. This is especially true if a destructive test is involved. So a small sample is taken from the population, using a good sampling plan, and the results of the sample are reported as their descriptive statistics. Next the sample data can be used in an inferential statistics that can make an estimate of what the parameters for the entire population. The measure of the sample data can be referred to as a *statistic*, and these are used to make an estimate about a value for the larger population, called a *parameter*. As seen in Chap. 2, the average for sample is the statistic called the "sample mean," and it is the best estimate of the larger parameter, the "population mean." As depicted in Fig. 1.1, if we were making an estimate of the center, the sample statistic would be the average for the sample and use that information with an inferential statistics to make estimate of the average for the entire batch or the population parameter. The sample statistics will be discussed in Chap. 2, and estimating a population parameter will be presented in Chap. 3.

Table 1.1 Results of random samples reported as percent label claim

Sample	%LC	Sample	%LC
1	100.102	11	100.416
2	98.724	12	100.945
3	99.745	13	97.112
4	98.675	14	99.655
5	100.763	15	102.814
6	101.734	16	94.667
7	99.145	17	100.322
8	98.453	18	101.369
9	102.217	19	99.349
10	98.234	20	98.153

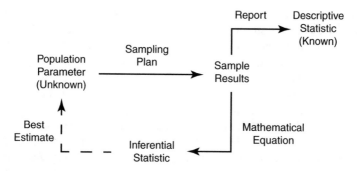

Fig. 1.1 Descriptive and inferential statistics

We will assume that the scientist is careful and competent; thus we can be 100% confident in his reporting of the descriptive statistic. However, when employing an inferential statistic and making a statement about the entire population without measuring every unit in that population, we have to live with the possibility that the estimate could be wrong. We can never be 100% confident in our decision; we have to allow for a certain amount of error. Controlling these possible errors will be discussed in Chap. 4.

1.2 Importance of Identifying the Variables in a Study

Identification of the types of variables involved in a study is critical to deciding what inferential statistic should be used to evaluate data. Equally important, certain variables can be controlled by the researcher, while others are beyond the researcher's control. A variety of names are associated with the labels that can be attached to these variables. As will be seen at the end of this section, identification of the types of variables that the researcher can and cannot control and whether these variables are qualitative or quantitative will dictate the most appropriate inferential statistic to use in evaluating research outcomes.

A synonym for a variable is a *factor*, a term which will be used in Chaps. 5 and 6.

1.2.1 Discrete Variables

For the purposes of this book, the term discrete variables represent *qualitative variables* or categories. Example would include experimental vs. control, batch A vs. batch B, or technician 1 vs. 2 vs. 3. A synonym would be *nominal variables* (the origin is the Latin word *nominalis* meaning "of a name") because a name, not a quantity, is associated with each level of the variable. There are two rules associated with discrete variables. First the categories must be *mutually exclusive*. For example, a single tablet or data point must come from either batch A or batch B; it cannot come from both, thus mutual exclusivity. Second, the levels of the discrete must be *exhaustive*. In this case there were no sample from batch C or D or any other batch. Thus, batch A and B exhausted all possibilities. In another example, a researcher is comparing three possible dissolution instruments where any result or data point must come from only one of three mutually exclusive instruments, and those three instruments would exhaust all the possible instruments in the study.

Sometimes, to meet the exhaustive criteria, an "other" category will be included to cover all possible outcomes. For example, possible a recording sheet for a flame test check boxes exists for red, yellow, or no color. What if a blue color is noted? The recording sheet is not exhaustive. An appropriate recording sheet would include red, yellow, no color, and others, creating an exhaustive list of possible outcomes.

1.2.2 Continuous Variables

Continuous variables are *quantitative variables*; examples would include length, weight, volume, or percent. These represent the most commonly encountered outcomes in pharmacy research and involve some type of measurement. With a discrete variable, outcomes or measures are clearly separated from one another (e.g., experimental vs. control group). With continuous variables it is possible to imagine more possible values between them. Theoretically, no matter how close two measures are together, a difference could be found if a more precise instrument were used. Consider weight, which is a continuous variable; it can be measured by kilograms, grams, milligrams, micrograms, or even nanograms. But assume that the precision of an older analytical balance is milligrams and the reading is 14 milligrams. With a more precise instrument, it would be possible to measure an infinite number of possible results between a half a unit above and a half a unit below the measure on the older balance (13.5 to 14.5 mg). Thus, with a continuous variable, there are no gaps or interruptions. There is also a relative positioning and consistency of measurements. For example, 28 mg is twice as large as the original 14 mg and 56 mg is another doubling in the magnitude of that weight.

It is possible to make a discrete variable from continuous variable by choosing selected values or dividing the variable into different categories. Occasionally, a continuous variable is presented on a *rating scale* or modified into a discrete variable. For example, study results may be (1) dichotomized either above or below the midpoint; (2) arbitrarily classified as high, medium, or low results; or (3) measured on a continuum that either "passes" or "fails" a predefined level. Even though each of these examples represents the results of a continuous measurement, by placing them a priori (before the test) on a rating scale, they can be handled as discrete variables. For example, temperature is a continuous variable, but assume in a study the water bath is controlled to be at a low temperature of 32 °C and high temperature 39 °C. In the case the researcher has replace a continuous variable with a discrete variable of low and high temperature.

A continuous variable could be converted to scale called an *interval scale*, where the difference between each level of the scale is equal. The scales represent a quantitative variable with equal differences between scale values; however, ratios between the scale values have no meaning because of an arbitrary zero. For example, the ratio between 40 °F and 20 °F does not imply that the former measure is twice as hot as the second. If a genuine zero is within an interval scale, it becomes a *ratio scale*, for example, measures of weight or height. If an object weighs 500 mg and a second object weighs 250 mg, the first object is twice the weight of the second. Other examples of ratio scales would include percentage scales and frequency counts. With interval and ratio scales, most arithmetic operations (e.g., addition and subtraction) are permissible with these numbers. Ratio and interval scales are sometimes referred to as *metric scales*.

1.2.3 Ordinal Variables

Ordinal variables are quantitative, but the magnitude of difference between consecutive units of measure is not necessarily equal. In the previous example, a doubling of 14 mg would produce 28 mg. With continuous variable there are consistent magnitudes of difference. With ordinal variables (from the Latin word *ordinalis* meaning "relating to order in a series"), there is an ordered ranking, but the intervals between values may not be the same. An example is a just about right or JAR scale (Kemp et al. 2009). Here volunteers are asked to evaluate the sweetness of an over-the-counter product with values: very sweet, moderately sweet, somewhat sweet, just about right, sour, moderately sour, and very sour. Granted this is a subjective scale, but the seven levels are listed in an order of sweetness. However, is the magnitude the same between very sweet to moderately sweet or between moderately sweet to sweet? Statistically these could be evaluated as seven discrete categories of sweetness or quantitatively as an ordinal scale. If an ordinal scale is involved, the preferred type of inferential statistics are the nonparametric tests which will be discussed in Chaps. 5 and 6.

Both nominal categories and ordinal scales are sometimes referred to as *nonmetric scales*. For both scales it is possible to have only two possible levels. These are termed dichotomous or binary variables. If there are no relative positions (i.e., technician 1 vs. technician2), it is a dichotomous nominal variable. If there is a relative position (e.g., passing or failing a criterion), the variable is a dichotomous ordinal scale. For simplicity, the term discrete variable will be used for these nonmetric scales.

1.2.4 Independent vs. Dependent Variable

In addition to a variable being defined as discrete or continuous (rarely ordinal), it may also be considered independent or dependent. Most statistical tests require one or more independent variables that are established in advance and controlled by the researcher. The *independent variable* (synonym – *predictor variable, predictor factor*) allows the researcher to control some of the study environment. A *dependent variable* (synonyms – *response variable, response factor* or *outcome*) is then measured against its independent counterpart(s). These dependent variables are beyond the researcher's control and dependent on the levels of the independent variable used in the study. Independent variables are usually qualitative (nominal) variables but also may be continuous or ordinal. Examples are provided in the next section.

In designing any research study, the investigator must control or remove as many extraneous variables as possible, measure the outcome of only the dependent variable, and compare these results based on the different levels or categories of the independent variable(s). The extraneous factors that might influence the dependent variable's results are known as *confounding variables* or *nuisance variables*. Careful protocol-driven study designs (or SOP), good sampling, and appropriate/careful scientific procedures can help remove some of these confounding variables. In the previous example (Table 1.1), using different instruments to measure the contents at different sites may produce different results even though the tablets are the same at all sites, because of potential differences in the measurement instruments. One approach would be to use an *experiment design* study to identify the one or two independent variables having the greatest impact on the outcome dependent variable and then design the research study to focus on those identified variables.

1.2.5 Using Variables to Identify Inferential Statistical Tests

The types of variables involved in a study (continuous or discrete) and identification of which can and cannot be controlled by the researcher will determine the most appropriate inferential statistical tests to use in evaluating study results. A flow chart is provided in Appendix A to help with this decision-making. Two examples follow that lead through Appendix A.

The first example involves samples that were taken from a specific batch of drug and randomly divided into two groups of tablets. One group was assayed by the manufacturer's own quality control laboratories. The second group of tablets was sent to a contract laboratory for identical analysis. The obvious question would be: Are there differences in the outcomes between the two laboratories? The researcher has designed the study and selected only two laboratories; therefore the independent variable is discrete with two levels (manufacturer's lab vs. contract lab). The result of the analysis at the two laboratories is the percent label claim; therefore the outcome (dependent variable) is a continuous, quantifiable variable. Using Appendix A, Panel A, the first question is: Is there an independent variable? Yes, proceed down.

Is the independent variable continuous or discrete? Discrete (laboratory – manufacture or contract), proceed down. Are the results based on proportions of success or failure? No, proceed down. Is the dependent variable continuous or discrete? Continuous (percent label claim), proceed to Panel B (for discrete independent variable and continuous dependent variable). How many independent variables are being tested? Only one (laboratory), proceed down. How many levels of the independent variable? Two, proceed down left side. Is the data paired or unpaired? This will be discussed in Chap. 5, but since data is collected from two independent laboratories on separate samples, it is considered unpaired. Then the last question relates to rule for parametric tests (Sect. 3.2). If the rules are met, then the appropriate test would be the two sample t-test (Chap. 5). If the rules are not met, a nonparametric alternative would be required.

For the second example, two different analytical methods are used to evaluate samples. The researcher selects sample products that might be assayed by these methods and tests the sample using both methods. The results (outcomes) are reported as percent. Notice in this case the only thing the researcher has control is the selection of the two analytical methods but cannot control results produced by each method. In this example there is no independent variable. The research questions would be – what is the relationship in results using the two methods? As one gets bigger will the second one also gets bigger, gets smaller or show no relationship to the first? Turning the Panel A in Appendix A, is there an independent variable? No, proceed to the right. Is the dependent variable continuous of discrete? Percent is continuous, proceed down and go to panel D for two or more continuous dependent variables. The next question is associated with the distribution of the data. Are the distributions somewhat symmetrical (symmetry will be discussed in Sect. 3.4)? If yes, proceed down to the parametric test; if not go with the nonparametric alternative. Assuming symmetry the inferential test of choice for determining the strength of this relationship would be correlation (Chap. 6).

1.3 Critical Assumptions for All Statistical Tests

Different statistics (descriptive or inferential) will have different rules and requirements, and these will be discussed as the tests are presented in the following chapters. However, there are two common sense assumptions that apply for all inferential tests. First, all data is collected using a good sampling plan. The sample must be representative of the population about which the final decision will be made. One cannot sample from batch A and make a statement about batch B. Similarly sample from the first half of a production run may not be representative of the second half. Thus, the sampling must occur throughout the entire run of the product.

The second assumption is that every data point is independent of any other data point in the sample. This involves good science with accurate and correctly calibrated equipment and clean and proper handling of sample. The 12th sample analyzed should not be effected by the 11th or any previous data point.

Good sampling and independent measurements fall outside the statistician's domain. They require good scientific practices on the part of the analyst or researcher collecting the data.

1.4   Steps in Performing an Inferential Statistical Test

Many individuals envision statistics as a labyrinth of numerical machinations. Thus, they are fearful of exploring the subject. The effective use of statistics requires more than knowledge of the required formulas and mathematics. This is especially true today, with computers that can quickly analyze sample data. This is one of the reasons that statistical equation and mathematical proofs are only supplemental to the text in the following chapters. However, there are several important steps to completing an appropriate statistical test.

Step 1. Establish a research question It is impossible to acquire new knowledge and to conduct research without a clear idea of what you wish to explore. Research is expensive and time-consuming. If volunteers are involved in a clinical trial, there are additional inherent risks for these individuals. Therefore clear objectives or research questions must be created before any data is collected. For example, a researcher wants to know if a tablet produced at three different manufacturing sites will have the same hardness. The research question simply stated: Is the tablet hardness for this product the same for the three manufacturing sites?

Step 2. Formulate a hypothesis Discussed in Chap. 4, the researcher should formulate a hypothesis that will be either rejected or fail to be rejected based on the results of the statistical test. In this case, the simple hypothesis that is being tested is that Facility A equals Facility B equals Facility C. The only alternative to this hypothesis is that the results for the three facilities are not all equal to each other. Hypotheses are specific for the different inferential statistics and will be presented in Chaps. 5 and 6.

Step 2. Select an appropriate test Using information about the data (identifying the dependent and independent variables), the correct test is selected based on whether these variables are discrete or continuous. For example, manufacturing sites A, B, and C represent an independent variable with three discrete levels and the result for the tablet hardness continuous variable (kp, kilopond) dependent upon the manufacturing site from which it was sampled. Therefore, the most appropriate statistical test would be one that can handle a continuous dependent variable and a discrete independent variable with three categories. Proceeding once again through Appendix A, the researcher would conclude that the "analysis of variance" test would be most appropriate (assuming the requirements for a parametric test that will be discussed in Chap. 3).

A common mistake is to collect the data first, without consideration of these first three requirements for statistical tests, only to realize that a statistical judgment cannot be made because of the arbitrary format of the data. The next two steps are relatively simple but require care on the part of the researcher.

Step 4. Sample correctly   Samples should be selected from each site using an appropriate sampling plan (below). An appropriate sample size should be determined to provide the most accurate results (Sect. 1.5).

Step 5. Collect data   Care should be taken during the hardness testing of samples to ensure that each observed result is independent of any other sample. If not automatically transferred from the instrument to a database, care should be taken for any transfer of information from a notebook to the computer.

Step 6. Perform statistical test   This is the only step in a statistical process that actually involves the mathematics associated with a statistical analysis. Many commercially available computer packages are available to save the tedium of detailed mathematical manipulations.

Step 7. Make a decision   This final step is the most important step. Based on the data collected and inferential statistical tests performed on the sample data, a statement (inference) is made regarding the entire population from which the sample was drawn. In this example, based on the results of the test statistics, the hypothesis that tablets will have the same hardness from all three manufacturing sites is rejected or results fail to reject the hypothesis. As discussed in Chap. 4, the initial hypothesis can be rejected, but can never be proven to be true.

1.5   Sampling Plans

Ideally samples should be drawn at random from the population being studied. A random sample means that every data point in the population has an equal chance of being selected for the study. Printed random number tables are available (e.g., De Muth, p. 688) and relatively easy to use. Random number generators can also be found on hand calculators or computers and even available as cell phone apps. As an example, assume that there are 30 vials periodically collected during a production run and numbered consecutively in the order they were sampled. These are sent to the quality control department, but only ten vials are needed to test for particulates (USP <788>). How does the analyst select 10 from the available 30 vials? The analyst used her cell phone app to randomly select vials 3, 4, 5, 7, 14, 15, 21, 23, 25, and 29. Note that during the process, 21 came up twice, so the second time it was eliminated and sampling continued until ten different unique numbers were selected. Using this process all 30 vials had an equal chance of being selected.

Random sampling is especially useful for selecting materials to be exposed to different test conditions (independent variable). For example, assume the 30 vials were instead to be analyzed for amount of active ingredient using 2 different methods (e.g., spectroscopy vs. chromatography). Assume that an equal number of vials (15) will be tested using each method. In this case one method is selected a priori (spectroscopy), and a random number generator is used to select 15 out of the 30 vials. These 15 vials would be tested using spectroscopy. The remaining 15 vials would be assayed by chromatography. Note that the method was selected before the random sampling, and again each vial had an equal chance of being assayed by one of the two methods.

The disadvantage with random sampling is that it may not be always practical. For example, trying to randomly sample from a production run of over one million units and the random number generator identified unit 198,432 which would be difficult or impossible to capture that exact unit. Thus, other selective sampling plans may be more appropriate.

Systematic sampling is a process by which every nth object is selected. In the pharmaceutical industry, this might be done during the production run of a certain tablet, where at selected time periods (every 30 or 60 minutes), tablets are randomly selected as they come off the tablet press and weighed to ensure the process is within control specifications. In this production example, the time selected during the hour can be randomly chosen in an attempt to detect any periodicity (regular pattern) in the production run.

In *stratified sampling* the population is divided into groups (strata) with similar characteristics, and then individuals or objects can be randomly selected from each group. For example, with blend uniformity during the mixing process, specific target locations may be part of the sampling plan where a thief is used to sample from the top, middle, bottom, or other locations in a blend. In a clinical trial, gender may be important, so males are randomly assigned to different treatment options, and females are randomly assigned separately, creating strata based on gender.

Also known as "multistage" sampling, *cluster sampling* is employed when there are many individual "primary" units that are clustered together in "secondary" larger units that can be subsampled. For example, individual tablets (primary) are contained in bottles (secondary) sampled at the end of a production run. Assume that 150 containers of a bulk powder chemical arrive at a pharmaceutical manufacturer and the quality control laboratory needs to sample these for the accuracy of the chemical or lack of contaminants. Rather than sampling each container, quality control randomly select ten containers. Then within each of the ten containers, they further extract random samples (from the top, middle, or bottom) to be assayed.

Any of these previous approaches to collect data are called *probabilistic sampling plans* and will increase the likelihood of collecting a good sample. There are a number of *non-probability sample plans*, where the samples are gathered by a process that does not give all the data in the population equal chances of being selected. One example is a sample of convenience, because samples are easily accessible to the researcher. Results for these non-probabilistic methods may be meaningless and should be avoided.

1.6   Reportable Values, Rounding, and Significant Figures

If performing a statistical analysis, rounding should never occur until a summary value is determined. Fortunately computer software will carry out procedures without rounding until the summary value is reported. This summary value is termed the *reportable value* and represents the end result of a completed measurement procedure, as documented (USP General Notices).

If an intermediate result is used for additional calculation (e.g., a sample average in an inferential statistic), the value should not be rounded. However, if those same intermediate results are used as part of the reporting procedure (e.g., sample average), it may be rounded in the report, but the original (not rounded) value should be used for any additional required calculations.

There are different rounding conventions, the most common is the one recommended in the USP General Notices "When rounding is required, consider only one digit in the decimal place to the right of the last place in the limit expression. If this digit is smaller than 5, it is eliminated and the preceding digit is unchanged. If this digit is equal to or greater than 5, it is eliminated and the preceding digit is increased by 1."

Reportable values involve summaries reported with a number of *significant figures*. Using computer software helps with the issue of retaining significant figures during calculations because there programs will retain as many digits as they are capable of holding throughout the computations. Most computer program will round the final reportable value to a predetermined metric. For reporting purposes the number of significant figures should be based on common sense. For example, if original data is measured to the ten of a milligram, the reporting of the average should be either to the tenth or one-hundredth of a milligram. Reporting the average to the whole milligram would not be informative enough, and reporting to the microgram level would be overkill. For counting purposes, (1) all non-zero digits are always significant, (2) any zeros between two significant digits are significant, and (3) only one final or trailing zero is significant.

Other authors have established more definitive rules for rounding and significant figures (Scott 1987; Torbeck 2004). The convention used in this book for reportable values will be to report one significant figure past the level of digits measured in the sample observations, if practical, or the value produced by software output.

1.7   Introduction to Minitab

Many commercial software packages are available for presenting descriptive statistics or doing inferential statistical analysis. They are usually easier, quicker, and more accurate than hand calculations. With easy access to computer software, the sixth step (Sect. 1.4) in a statistical procedures (perform statistical test) may be the least important component of a statistical procedure. However, commercial software can give the user a false sense of security, and it is important to understand the software and how to enter and query the data. The availability of software makes the task easier but does not eliminate the need for a good understanding of basic statistics and which test is appropriate for the given situation. Even using sophisticated packages, the researcher still needs to interpret the output and determine what the results mean with respect to the analysis performed.

No software package is "perfect," and before purchasing any package, the potential users should determine the types of applications they commonly require, access a demonstration version of the software, and determine if it meets their needs. One available statistical package is available in Excel® by Microsoft. Readily available it offers an inexpensive method to analyze data using some of the most commonly used inferential statistics. A stepwise approach to these Excel tests are available (Billo 2001; De Muth 2014). Many other commercial software packages are available, including JMP, SAS, SPSS, Statgraphics, and Systat.

This book will focus on Minitab 19® by Minitab LLC.[1] The author selected Minitab as the software of choice because it is efficient for handling sample data and offers a comprehensive statistical package. It has a user-friendly interface, reasonably price, and multiple free online teaching resources. The author commonly uses this software for on-site industrial short courses and has used versions of Minitab since the early 1970s when the software was run on mainframe computers with punch cards!

1.7.1 Why Minitab?

Minitab is an easily understood, comprehensive statistical software packages. Many pharmaceutical manufacturers already use different versions of Minitab. The current version of this software was released in June 2019 (right the middle of writing this book!). Fortunately most of the changes involve the processing speed, design of the worksheet and session window, and applications not covered in this book. Commands for accessing statistical tests are the same as those in Minitab 17, and Minitab 18 has not changed since Minitab 17; so commands and tests results in this book will be similar for at least the last three most recent versions of Minitab.

Worksheets are similar to Excel and it is easy to enter data, load, copy, or paste data from other sources. The dropdown menus are well organized and complement the descriptive terms used in this book. The dialog boxes are straightforward and easy to use and select program options. Various graphs can help researchers visualize and explore sample data. Output results can be controlled for more or less information and easily copied for transfer to report documents.

---

[1] MINITAB® and all other trademarks and logos for the company's products and services are the exclusive property of Minitab, LLC. All other marks referenced remain the property of their respective owners. See minitab.com for more information.

The statistical applications covered in this book are only the tip of the iceberg for the applications available in Minitab. In addition to basic statistics covered in this book, Minitab 19 can perform (1) measurement systems analysis, (2) process mapping, (3) process improvement tools, (4) control charts, (5) design of experiments, and (6) process capability measures.

Traditionally, Minitab was available only for a Windows platform. With Minitab 19 it also became available for MAC computers.

Additional resources and training are available including for Minitab users: Quality Trainer®; Minitab Express™; Companion by Minitab®; Mentoring by Minitab®; Technical Support by Minitab™; and Training by Minitab™ (www. minitab.com/en-us/).

1.7.2 Getting Started with Minitab

The work areas for Minitab 19 contain three sections (Fig. 1.2): (1) three upper right screens that show the results of analyses (Sessions window), (2) the columns and rows in the lower right that contain the data (Worksheet window), and (3) the Navigator window to the left that lists most recent activities in the project. Previous versions of Minitab included only the sessions and worksheet window, and the navigator was something that could be recovered if needed. The Navigator was added to allow for quicker access to previous tests, graphs, and results. By clicking on the desired line on the Navigator window, the report is presented in the Sessions window, saving time scrolling up and down in the report.

With Minitab most of the other statistical operations will involve pointing and clicking on the "Stats" and "Graph" commands on the application title bar (Fig. 1.3). Throughout this book, the procedures for initiating the tests in Minitab will be discussed, and output will be displayed to determining how to make decision results from inferential test results. Stepwise progression to initiate the test will be presented as:

Title bar → Drop down choice 1 → Drop down choice 2 → Main dialog box

For example, accessing a two-sample test (Sect. 5.3), the initial steps would be:

Stats → Basic Statistics → 2 sample t

The main dialog box for this test appears in Fig. 1.4. The selections for graphics, changing test limits, or other special features/tests will be presented as optional choices in the main dialog box and lead to other dialog boxes. For every statistical test discussed in this book and available in Minitab, there will be a detailed discussion of the steps to initiate the procedure, options available, the information of the output report, and how to interpret the results.

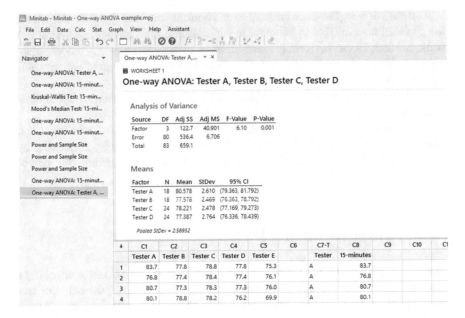

Fig. 1.2 Opening screen for Minitab 19

Fig. 1.3 Title bar for Minitab

In the Worksheet window is similar to Excel. Data is normally entered with each row representing an observation/sample/person and each column a variable. An example is presented in Fig. 1.5. Notice there are three types of variables: (1) quantitative continuous variables have no extension after the column number; (2) categorical discrete variables have a "T" extension; and (3) for dates the column extension is "D." In this case continuous variables are in columns 5, 7, 8, and 9, while discrete variables are in columns 3, 4, and 6. Dates the samples were collected are in column 2.

Minitab allows data to be also arranged such that each column would represent a level of the dependent variable (similar to data analysis with Excel). Certain test offer the option of using data arranged by levels in different column. Also, several test can accept summary data (sample size, sample average, and sample dispersion). These will be discussed in Chaps. 5 and 6.

In the main dialog box (Fig. 1.4), a list of variables available for analysis will appear in the box to the left. The variables to be used in the test are selected by

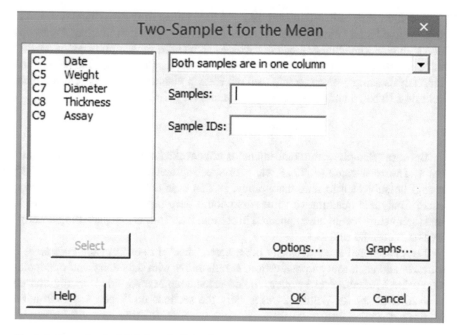

Fig. 1.4 Example of a Minitab main dialog box

↓	C1-T	C2-D	C3-T	C4-T	C5	C6-T	C7	C8	C9
	Sample	Date	Technician	Balance	Weight	Caliper	Diameter	Thickness	Assay
1	TT-01	2/15/2019	AJ	A	95.3	X	11.2	6.1	100.2
2	TT-02	2/15/2019	LC	B	93.2	X	11.1	6.0	99.8
3	TT-03	2/15/2019	DD	C	94.6	X	11.2	6.0	97.3
4	TT-04	2/16/2019	AJ	C	95.0	Y	11.0	5.9	99.5
5	TT-05	2/16/2019	DD	B	94.8	Y	11.3	6.2	101.4
6	TT-06	2/16/2019	LC	A	95.2	Y	11.4	6.1	100.4
7	TT-07	2/17/2019	AJ	C	96.1	X	10.9	6.0	99.7

Fig. 1.5 Example of a Minitab worksheet

double-clicking the variable name. For example, assume that there were two "Assay" methods and tests were performed on "Weight" comparing the two methods. The cursor would be placed on "Samples" box and the variable "Weight" double-clicked to select the dependent continuous variable. The cursor would be placed on "Sample IDs" box and, "Assay" would be double-clicked as the independent discrete variable. Once these are selected, the test would proceed, and options or graphs could be selected (as detailed in Sect. 5.3.4.1). The variable columns that are labeled as "text" can only be selected for options requiring discrete variables, and numerical column can only be selected for options requiring continuous variables.

Another example involves sampling. As noted most of the statistical operations will involve "Stats" and "Graph" commands on the application title bar. An exception is the random number generator which can be used for sampling and is located in "Calc" on the title bar. Using the previous example, 15 vials need to be selected for analysis using a spectroscopy method. Fifteen vials are randomly selected from column 1 (labeled vials) and place in column 2 (Fig. 1.6).

Calc → Random data → Select from column

Because "Sample with replacement" is not checked in the dialog box, 15 different vial were selected (3, 12, 8, etc.) for spectrographic assay. The remaining 15 would be subjected to chromatography. In this case only the vial numbers were used. Analytical quantitative values also could have been presented in column 1, and their values would have appeared in column 2 for further analysis using descriptive or inferential statistics.

Data is entered much like any other spreadsheet. In general each column is a variable and each row an observation. If unfamiliar with data entry and manipulation using Minitab, please explore the downloadable user's guide "Getting Started with Minitab 19 for Windows" especially the sections on "Open and examine a worksheet" (Minitab).

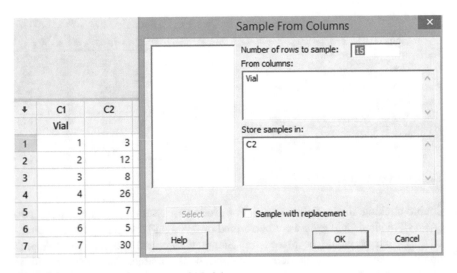

Fig. 1.6 Results of a random sample of 15 vials

1.7.3 Minitab Statistical Procedures Detailed in the Following Chapters

For most of the statistical tests described in this book, there will be a standard template describing the how to use the test, available options, and ways to interpret the results (e.g., the one-way ANOVA in Sect. 5.5.5.1):

Procedure	The steps to initiate the statistical test
Data input	Which variables are selected and where they are entered into the main dialog box (procedure described in previous section)
Options, graphs, comparisons, model, results, and/or storage	changes to the default commands to control the outcome and reports generated
Report	The results reported by the Minitab procedure
Interpretation	How to determine if the results are statistically significant or not

Hopefully these guides will help the reader to more effectively use Minitab. Other references for the use of earlier versions of Minitab include Ryan and Joiner (2002), Lesik (2010), and Khan (2013).

References

Billo J (2001) Excel for chemists: a comprehensive guide, 2nd edn. Wiley-VCH, New York

De Muth JE (2014) Basic statistics and pharmaceutical statistical applications, 3rd edn. CRC Press, Boca Raton, FL

Kemp SE, Hollowood T, Hort J (2009) Sensory evaluation: a practical handbook. Wiley, Oxford, UK, pp 137–138

Khan RM (2013) Problem solving and data analysis using Minitab. Wiley, Southern Gate, UK

Lesik SA (2010) Applied statistical inference with MINITAB. CRC Press, Boca Raton

Minitab Getting Started. https://www.minitab.com/uploadedFiles/Documents/getting-started/MinitabGettingStarted_EN.pdf

Ryan BF, Joiner BL (2002) MINITAB handbook: update for release 16. Brooks/Cole, Boston

Scott S (1987) Rules for propagation of significant figures. J Chem Ed 64(5):471

Torbeck LD (2004) Significant digits and rounding. Pharmacopeial Forum 30(3):1090–1095

USP 42-NF 37 (2019a) General chapter <788> particulate matter in injection. US Pharmacopeial Convention, Rockville

USP 42-NF 37 (2019b) General notices 7.20 Rounding rules. US Pharmacopeial Convention, Rockville

Chapter 2
Descriptive Statistics and Univariate Analysis

Abstract Results from an experiment will create numerous data points. The organization and summary of these data are termed descriptive statistics. This chapter presents the various ways to report descriptive statistics as numerical text and/or graphics. For qualitative (categorical) data, the use of tables, pie charts, and bar charts are the most appropriate ways to summarize the information. With quantitative (measurable) data, the researcher is interested in reporting both the center of the samples and the dispersion of data points around that center. Histograms, dot plots, and box-and-whisker plots are appropriate graphics for quantitative data. These descriptive statistics provide the information to be used for the inferential statistics discussed in later chapters. A univariate statistics involves the analysis of a single variable, whereas a multivariate statistic evaluates the differences, relationships, or equivalence for a dependent variable based on levels of an associated independent variable in the study design.

Keywords Continuous variables · Descriptive statistics · Discrete variables · Measures of center · Measures of dispersion · Relative standard deviation · Standard deviation · Variance · Mean · Median

As seen in Fig. 1.1, any descriptive or inferential statistic required the collection and analysis of sample data that is representative of the population from which it was taken. The results of the data collection will be numerous *data points*. The role of descriptive statistics is to organize and summarize these data points and present the results in a more palatable manner to the person reviewing the outcomes of a study.

For the purposes of this book, the term *univariate* refers to analysis of a single variable. For example, a univariate result might be summary of results expressed as percent, even though data was collected under several different test conditions. These are usually descriptive statistics but may involve inferential tests (Sects. 3.6 and 5.2). *Multivariate* would be reserved for tests where results are evaluated under the multiple conditions (variables) and possible differences, similarities, or relationships between the variables (Chaps. 5, 6, and 7). Using the same example, the percent would be compared for each of the different levels of a second variable.

© American Association of Pharmaceutical Scientists 2019
J. E. De Muth, *Practical Statistics for Pharmaceutical Analysis*, AAPS
Advances in the Pharmaceutical Sciences Series 40,
https://doi.org/10.1007/978-3-030-33989-0_2

Table 2.1 Table summary of the analysis of tablets labeled 50 mg

mg of drug	Frequency (n)	Percent	Cumulative n	Cumulative %
≤ 45.4	3	2.5	3	2.5
45.5–49.4	23	19.2	26	21.7
49.5–50.5	57	47.5	83	69.2
50.6–55.5	31	25.8	114	95.0
≥ 55.6	6	5.0	120	100.0
Total	120	100		

2.1   Reporting Results for Discrete Sample Data

For discrete variables (*categorical variables*), with discrete levels of results, the reporting is fairly simple. Numerically the results are presented as tables listing the frequency count and percent of responses in each level or category of the discrete variable. In Table 2.1 there are 120 assay results for tablets that are labeled as 50 mg. It is possible to take quantitative data and break it into discrete categories. In this case the results are broken down into five categories (Table 2.1). Additional information is also presented in this table since the categories are in an ordinal arrangement by magnitude of dosage, with values increasing going down the table. In this case it is possible also to report the cumulative frequency and the cumulative percent for each level of this discrete variable. For example, 69.2% of the results are less than or equal to 50.5 mg. Also notice that the five categories are mutually exclusive and exhaustive. Each result can only fall into one category, and all possible results are covered by the five categories.

Visually, bar charts and pie charts are used to display results for discrete variables. Figure 2.1 represents the same information presented in Table 2.1, but in graphic formats. Notice the clear separation between the five bars in the bar chart indicating five discrete levels. The spacing is different when presenting continuous data.

2.2   Reporting Results for Continuous Sample Data

For continuous data, because it is quantitative, increased information is available that can be summarized and presented to the reader. Since data are on a continuum, the first point of interest is the location of the center for any particular sample. But of equal importance is how the sample data vary around that center (the dispersion of the data). There are different ways to measure the center and dispersion of sample data, as well as new graphics for visualizing these results.

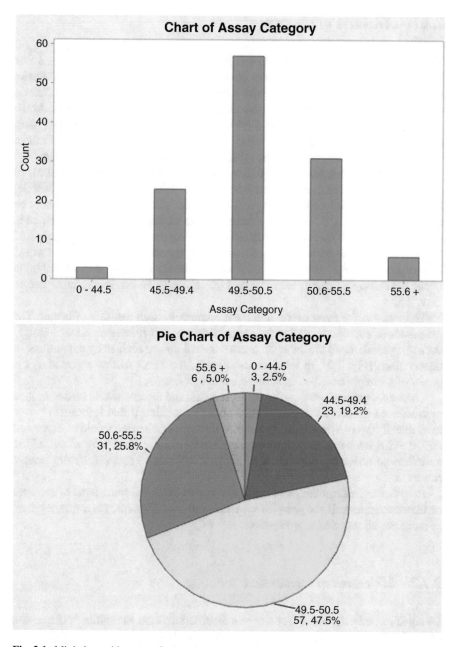

Fig. 2.1 Minitab graphic output for bar chart and pie chart for data in Table 2.1

2.2.1 Measures of Central

There are three measure of central tendency – the mode, median, and mean. *Mode* is the simplest measure of center and merely represents the most frequent occurrence(s). It is possible to not have a mode in the case where every data point is unique and there are no duplicate values (all frequency counts equal one). At the same time, there could be multiple modes where several values have an equal number of multiple data points. The mode is rarely reported and is usually associated with reporting the category with the greatest frequency for a discrete variable.

The *median* is that point on a continuous distribution where half of all the data fall below and half fall above the median, also called the fiftieth percentile (50%ile) it divides the data into equal groups. It is a robust statistics because it cannot be affected by an outlier (Chap. 8). To determine the median, the data are arranged in a sequential order. If there are an odd number of data points, the center value would be the median. If an even number of data points, the center two data points would be averaged. Both the mode and median are calculated by looking at the distribution of the data using a histogram or dot plot (Sect. 2.2.3). Examples are presented in Sect. 2.5.

The *mean* is the most useful and most commonly used measure of center. The magnitude of each value is taken into consideration. This *arithmetic mean* is simply the *average*, where all the data points are summed up and divided by the number of observations (Fig. 2.2). In most sample results, the mean will be reported as the center of a sample distribution and expressed as x-bar ($\bar{X}$).

These measures of center – the mode, median, and mean – would be determined or calculated the same for sample data or for population if that information were available. If it were a population parameter, the population mean would be expressed with the Greek letter μ. As will be seen in the next chapter (Sect. 3.6), when making an inference about the population μ, the best estimate to use will be the sample results $\bar{X}$.

Results for these measures of center are expressed in the same units of measure as the observations. If the samples are measured in milligrams, the mode, median, or mean are all reported as milligrams.

2.2.2 Measures of Dispersion

Of equal importance with the center of a distribution is the variability or dispersion of the data around that center. Like center of a distribution, there are also three measures of dispersion – range, variance, and standard deviations. With all three measures, the larger the value, the greater the dispersion.

The *range* is the simplest to calculate and represents the distance between the largest and smallest measure in the sample or population data set. The range may be useful for reporting the distance between the extreme values in a sample, but is not used in the calculation of any common inferential statistic.

	Sample Statistic	Population Parameter

Mean:

$$\bar{X} = \frac{\sum_{i=1}^{n} x_i}{n} \qquad \mu = \frac{\sum_{i=1}^{N} x_i}{N}$$

Variance:

$$S^2 = \frac{\sum_{i=1}^{n}(x_i - \bar{X})^2}{n-1} \qquad \sigma^2 = \frac{\sum_{i-1}^{N}(x_i - \mu)^2}{N}$$

Standard deviation:

$$S = \sqrt{S^2} \qquad \sigma = \sqrt{\sigma^2}$$

Relative standard deviation:

$$\%RSD = \frac{S}{\bar{X}} \times 100\%$$

Geometric mean:

$$\bar{X}_G = antilog\left(\frac{\sum \log(x_i)}{n}\right)$$

Geometric standard deviation:

$$S_G = exp\sqrt{\frac{\sum_{x=1}^{n}\left(\ln x_i / \bar{X}_G\right)^2}{n}}$$

Where: Σ represents sum all values from the first to n or N; n is the sample size; N is the size of the population; x_i are the individual data points

Fig. 2.2 Equations for measures of center and dispersion

A better measure would be to determine how each data point varies from the center. This could be simply determined by subtracting the mean from each data point and averaging the results. Unfortunately, with the mean as the center, the sum of all differences would always be zero (equal amount of negative and positive differences). To correct this each difference is squared (negative differences become positive), and these values are averages. The term used for the result is the *variance* which is the average of the squared deviation for all the data points (Fig. 2.2). The variance has no associated unit of measure. For example, with weights of tables, a variance expressed as mg^2 would make no sense. However, it would be desirable to have a measure of dispersion in the same units as the mean, thus the *standard deviation*. The standard deviation is the square root of the variance (Fig. 2.2) and is expressed in the same units as the mean. This standard deviation is sometime referred to as an *absolute standard deviation*. Coupled together mean and the standard deviation are the most commonly reported results for a continuous variable (mean ± standard deviation) and expressed in the same units of measure.

A word of caution, when one reports or sees reports for the mean and standard deviation, make sure that the author labels what is to the right of the ± sign. It is assumed to be the standard deviation, but if unlabeled it could be something else (Sect. 3.5).

Sorry, a little mathematical logic here. Notice for the variance in Fig. 2.2, the denominators for the population parameter and the sample statistic are different. If data were known about the entire population, the sum of the squared deviation would be divided by N number of observation in the population (as simple averaging). This would be labeled with the Greek letter sigma squared (σ^2) representing a population parameter. Usually it is not possible to sample an entire population, so a sample is collected form that population. The best estimate of population center (μ) would be the sample mean ($\bar{X}$). Similarly, the best estimate of the population variance (σ^2) would be the sample variance (S^2). This is an estimate, so the denominator is (n-1) referred to the *degrees of freedom*. There are different ways to define degrees of freedom (Walker 1940; Janson et al. 2015). The simplest is that degrees of freedom give a slightly more conservative result. One does not know the population sigma, the sample results will be slightly larger to compensate for this lack of knowledge. The squared deviations divided by $n-1$ will always be larger than the same value divided by n. Therefore the sample variance (S^2) will always be a larger value.

As seen in Fig. 2.2, both variance terms represent squared deviations, so both variance terms need to be square rooted to product a population standard deviation (σ) or a sample standard deviation (S). Sample results will be slightly larger and a more conservative estimate that the population standard deviation. Computer software will usually calculate sample variance and standard deviation rather than population results. A simple test to determine if the software or calculator is reporting a sample or population standard deviation is to enter the numbers 1, 2, and 3. If the result is 1.000, it is a sample standard deviation; if 0.815 it is a population standard deviation.

2.2.3 Relative Standard Deviation

The sample standard deviation is an absolute standard deviation. The variability of data may often be better described as a relative variation rather than as an absolute term. This can be accomplished by calculating the *coefficient of variation* (CV) that is the ratio of the standard deviation to the mean. This CV is usually multiplied by 100 (Fig. 2.2) and expressed as a percentage (*relative standard deviation*, RSD or %RSD). The RSD can be useful in many instances because it places variability in perspective to the distribution center. Thus, relative standard deviations present an additional method of expressing this variability, which takes into account its relative

Table 2.2 Example of absolute vs. relative standard deviations

	30.5	305	3050
	29.6	296	2960
	31.3	313	3130
	30.7	307	3070
	29.6	296	2960
	28.9	289	2890
	30.2	302	3020
	29.5	295	2950
	29.4	294	2940
	30.6	306	3060
Mean =	30.03	300.3	3003
S.D. =	0.742	7.424	74.244
%RSD =	2.472	2.472	2.472

magnitude. To illustrate this, Table 2.2 contains test results in the first column (the mean, standard deviation, and relative standard deviation are reported). In the second and third columns, initial values are presented as 10- and 100-fold increases. These increases also result in a 10- and 100-fold increase in both the mean and standard deviation, but the relative standard deviation remains constant. In the pharmaceutical industry, this can be used as a measure of precision between various batches of a drug, if measures are based on percent label claim. Many compendia criteria were based on the relative standard deviation rather than the absolute standard deviation. For example, liquid products can have different viscosities (presumably the greater the mean viscosity, the greater its standard deviation). But USP required the RSD to be not more than 2% regardless of the mean viscosity (USP <911>).

2.2.4   Graphics for Continuous Data

There are several types of graphics commonly used for continuous data. Histograms, dot plots, stem-and-leaf plots and box-and-whisker plots can be used for univariate analysis and scatter plots when comparing more than one continuous variable. The data presented in Table 2.1 represent five categories but are quantities that can be measured and also could be presented as continuous data (Table 2.3).

Histograms break the data into equally spaced *class intervals*. A histogram for the data in the previous table is presented in Fig. 2.3. Notice in contrast to a bar chart, there are no gaps or interruptions between the class intervals, representing a continuum of data. By default Minitab will establish the number of class intervals base on the sample size. An alternative method is to use Sturges' rule for determining the number of these class intervals (Sturges 1926).

A *dot plot* is very similar to a histogram where dots instead of bars are used to indicate the frequency of occurrences (Fig. 2.3). The *stem-and-leaf plot* is another

Table 2.3 Raw data of the analysis of tablets labeled 50 mg

49.5	50.1	55.0	45.2	49.8	52.2	50.0	49.9
46.6	46.8	53.1	49.8	50.5	49.8	50.9	49.9
50.0	52.0	53.2	59.2	48.9	54.1	48.1	50.0
50.9	50.0	48.6	50.2	49.0	47.9	50.1	48.5
49.9	53.6	49.7	55.9	55.3	50.7	51.4	50.4
50.3	41.3	50.3	47.2	50.1	50.8	51.6	51.3
50.1	50.3	55.8	55.4	51.0	50.0	49.1	49.4
50.6	54.3	49.6	52.7	51.1	52.5	49.3	49.5
50.7	50.1	50.2	58.1	47.2	45.9	49.3	50.0
49.9	49.5	45.8	60.1	47.5	50.3	50.3	50.6
50.4	49.6	49.7	45.6	47.8	50.4	51.2	46.5
50.0	49.6	49.9	49.8	53.9	51.2	50.2	49.6
51.8	50.0	48.6	50.0	50.2	49.9	50.0	49.7
44.2	47.1	50.0	54.7	50.6	50.5	46.3	49.8
50.5	49.7	50.4	57.3	49.6	51.9	49.8	50.2

visual presentation for continuous data. Also referred to as a *stemplot*, it contains features common to both the histogram and dot plot. Digits, instead of bars, are used to illustrate the spread and shape of the distribution. Each piece of data is divided into "leading" and "trailing" digits. All the leading digits are sorted from lowest to highest and listed to the left of a vertical line. These digits become the stem. The result looks like a histogram that has been rotated 90 degrees. One additional graphic for continuous data is the box-and-whisker plot which will be discussed in Sect. 2.3.2.

Some inferential tests involving scatter plots for showing relationships between more than one continuous variable will be discussed in Sect. 6.1.3.2.

2.3 Dealing with Asymmetrical Populations

As will be discussed is Sect. 3.2, the shape of the distribution for the population of a continuous variable is important. Most of the common tests require that the population has a *symmetrical distribution* (normal distribution). In cases where the sample mean and sample standard deviation are reported, the author is assuming a normal distribution in the population. One quick estimate is to compare the sample median and mean; if they are fairly close together, a normal distribution for the population can be assumed. Tests for normality will be discussed in Sect. 3.4.2. Plotting sample data using a histogram or dot plot may be useful (which is the best guess of the population). However, since sample sizes are usually small, the sample plots may not look anything like a symmetrical distribution. In these case alternative measures, graphics and inferential statistical tests are available.

Variable	N	Mean	StDev	Variance	CoefVar	Median	Range	Mode	N for Mode
Assay Results	120	50.367	2.723	7.415	5.41	50.000	18.800	50	11

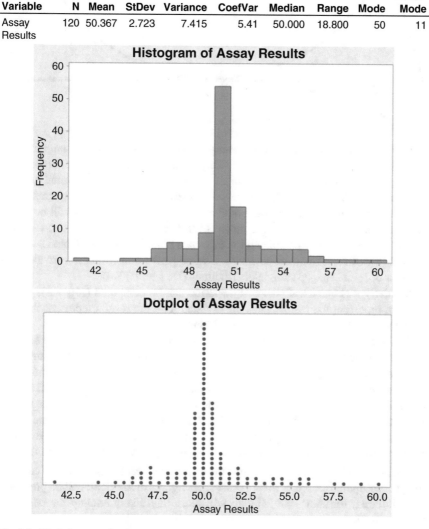

Fig. 2.3 Minitab output for the raw data that created Table 2.3

2.3.1   Geometric Mean and Standard Deviation

One of the most common symmetrical distributions is a positively skewed distribution with the tailing off or extending to the right (Fig. 2.4). This measure of skew will be discussed in the next chapter (Sect. 3.4.1), but at this point, realize that extreme values to the right will move the arithmetic mean more to the right and give a much higher value than would occur in a normal distribution. To present a center in a skewed distribution, either the median is reported or the geometric mean. The

Positive Skew   Negative Skew

Fig. 2.4 Examples of skewed distributions

geometric mean is calculated by averaging the log10 values for each data point. The logarithmic values are averaged, the average converted to the antilog and reported as the geometrical mean. The logarithmic values will create a more normal distribution than the original data. This process is sometime referred to as log transformation and the results as a log transformed mean. The geometric mean will be closed to the median than the arithmetic mean. In the case of a negatively skewed data, reciprocal values can be calculated for each data point and then evaluated using the geometric mean and standard deviation. Other types of transformation are also possible for different shaped sample distributions (De Muth 2014).

Similarly a geometrical standard deviation can be computed (Fig. 2.2), but the scientist reporting the results must clearly state that the results represent the geometric mean and geometric standard deviation.

As mentioned this book will not deal with formulas in the body of the text, but provide supplemental figures with the calculations and worked out examples in Appendix D. It is useful to note at this point that there are different types of formula that can be used to calculate the same results. In most cases there are *definitional formulas* which define the process used to calculate the results, for example, the equation for the geometrical mean in Fig. 2.2. However, definitional formula may be difficult to write for computer software, so alternative and equivalent formulas are used, called *computational formulas*. For example, the computational formula $\sqrt[n]{product\ of\ all\ data\ points}$ will give the exact same results and be much easier to code for computer software. In most cases, the figures in this book will be presented as definitional formulas, but Minitab will more than likely employ the computational equivalent.

2.3.2   Interquartile Range

If the population is not normally distributed, the mean and standard deviation are not appropriate measures of center and distribution. Instead the median and quartiles should be used. As noted above, the median is the 50%tile (Q_2) where half of all the observations fall above and half fall below. The 25%tile (Q_1) marks where 25% of the observations fall below, and 75%tile (Q_3) indicates where 25% of the observations fall above. A common measure of dispersion around the median is the

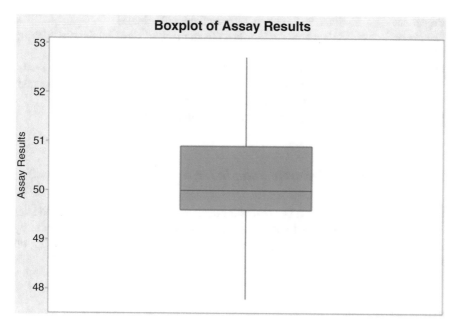

Fig. 2.5 Minitab graphic of a box-and-whisker plot representing data in Table 2.3

interquartile range or *IQR* (75%ile – 25%ile or $Q_3 - Q_1$). Graphically this is illustrated by a *box-and-whisker plot* illustrated in Fig. 2.5 for the data in Table 2.3. The interquartile range is represented by the box and the whiskers extending from the box to the extreme values. If the sample is close to a normal distribution, both halves of the box will be approximately equal in size, and the length of the whiskers should be similar. As the sample becomes more skewed, the differences in size for the two box halves will become greater. In the case of Fig. 2.5, the two halves of the box for the IRQ are not equal indicating a possible skew in the data (Chap. 3).

2.4 Minitab Applications for Descriptive Statistics

Minitab has applications for descriptive statistics as well as associated graphics. Steps vary depending on whether the variable is discrete or continuous. The following are procedures and results for various Minitab applications.

2.4.1 Tables for Discrete Sample Data

Tables can be created reporting the frequency count, percent, cumulative count, and/ or cumulative percent for each discrete variable selected.

Procedure	Stats ➜ Tables ➜ Tally Individual Variables.
Data input	Select column (left box) from the worksheet by double-clicking on each variable.
Options	Check the output required – counts, percents, cumulative counts, and/or cumulative percents. If the "store results" are checked, the tally results will appear in the next available columns.
Report	Create a listing for each column selected.

2.4.2 Graphs for Discrete Sample Data

Bar charts and pie charts can be created for discrete (categorical) variables and later modified (colors and labels) based on the need of the person preparing the report by double-clicking on the graph to edit and highlighting items and right clicking on the mouse.

2.4.2.1 Bar Chart

Procedure	Graph ➜ Bar Chart.
Initial decision	Select the type of chart required (simple, cluster or stacked).
Data input	Choose from the graph options and select the categorical variables (left box) from the worksheet.
Options	Multiple options are available for design and labeling of bar chart.
Report	Create a bar chart for each column selected.

2.4.2.2 Pie Chart

Procedure	Graph ➜ Pie Chart.
Data input	Choose from the graph options and select the categorical variables (left box) from the worksheet.
Options	Multiple options are available for design and labeling of pie chart.
Report	Create a pie chart for each column selected.

As noted previously, bar and pie charts and the reported images can be edited after they are created for content and color by highlighting the various sections and making the required changes.

2.4.3   Reporting for Continuous Sample Data

For continuous data various measures of center and dispersion can be selected and reported on description statistics for single or multiple variables.

Procedure	Stats → Basic Statistics → Display Descriptive Statistics.
Data input	Select one or multiple continuous variables (columns in left box) from the worksheet.
Statistics	Select from a variety of options for measures of center and dispersion. Descriptions of these options are listed in Table 2.4.
Graph	By default there are no graphics. Graphic can be selected to create histograms (with or without a normal curve), plot of individual values (similar to a dot plot), or a box plot.
Report	Create a listing of selected measure for each column selected. Optional graphics are possible if selected.

Table 2.4 Minitab options for descriptive statistics

Mean	Arithmetic mean
SE of mean	Standard error discussed in Chap. 3
Standard deviation	Sample standard deviation
Variance	Sample variance
Coefficient of variation	Relative standard deviation (%RSD)
First quartile	25%ile
Median	Sample median
Third quartile	27%ile
Interquartile range	IQR
Mode	Sample mode
Trimmed mean	Sample mean with 5% of largest and 5% of smallest values removed
Sum	Sum of x_i
Minimum	Smallest value
Maximum	Largest value
Range	Sample range
Sum of squares	Sum of x_i^2
Skewness	Measure of skew discussed in Chap. 3
Kurtosis	Measure of kurtosis discussed in Chap. 3
MSSD	Mean of the squared successive differences
N nonmissing	Sample size n
N missing	Number of missing data points
N total	Sum of N missing and N nonmissing
Cumulative N	Same as N nonmissing
Percent	Percent of N nonmissing
Cumulative percent	Cumulative percent of N nonmissing
Default	Displays default values
None	Removes all possible values
All	Displays all the possible descriptive values
Default values	

2.4.4 Graphics for Continuous Sample Data

Once created, histograms and dot plots can be modified (colors and labels) as needed by double-clicking on the graph to edit and highlighting items and right clicking on the mouse.

2.4.4.1 Histogram

Procedure	Graph ➜ Histogram.
Initial decision	Select the type of chart required (simple, with a fitted normal curve superimposed, with multiple groups or fitted with multiple groups).
Data input	Choose from the graph options and then select column (left box) from the worksheet.
Options	Multiple options are available for design and labeling of the histogram.
Report	Create a histogram for each column selected.

2.4.4.2 Dot Plot

Procedure	Graph ➜ Dotplot.
Initial decision	Select the type of chart required for a single variable or two variables (simple, with groups, or stacked).
Data input	Choose from the graph options and then select column (left box) from the worksheet.
Options	Multiple options are available for design and labeling of the dot plot.
Report	Create a dot plot for each column selected.

2.4.4.3 Stem-and-Leaf Plot

Procedure	Graph ➜ Stem-and-leaf.
Data input	Choose from the graph options and then select column (left box) from the worksheet.
Options	Multiple options are available for design and labeling of the stem-and-leaf plot including the removal of potential outliers (Chap. 7).
Report	Create a stem-and-leaf plot for each column selected.

2.4.4.4 Box-and-Whisker Plot

Procedure	Graph ➔ Boxplot.
Initial decision	Select the type of chart required for a single variable or two variables (simple or with groups).
Data input	Choose from the graph options and then select column (left box) from the worksheet.
Options	Multiple options are available for design and labeling of the box-and-whisker plot.
Report	Create a box-and-whisker plot for each column selected.

2.5 Examples

Look once again at the raw data presented in Table 2.3. It simply presents 120 data points in the order in which they were analyzed. Assume the researcher wants to evaluate the data as a discrete variable with five categories for the 50 mg. tablets (Table 2.1). A bar chart and pie chart can be created by Minitab for these five discrete categories which has already been presented (Fig. 2.1). The Minitab tabular output for these assays is presented in Fig. 2.6 which is identical to the hand-calculated results in Table 2.1. However, milligram is a quantifiable measurement, so descriptive statistics based on the milligrams of drug might be more informative than the five categories. Using the same information and going to descriptive statistics under "Stats" would create the information presented at the top in Fig. 2.3. Here the options selected were the three measures of center, three measures of dispersion, and the relative standard deviation (*CoefVar*). Note that the mode was 50.0 which had the greatest frequency of 11 data points. For reporting purposes the researcher would most likely report the mean and standard deviation (50.37 ± 2.72 mg.). Possibly the %RSD of 5.41% also would be reported along with the mean and standard deviation. The histogram and dot plot for the sample data are presented in the lower portion in Fig. 2.3.

For a second example of a continuous variable, samples are taken and evaluated to meet the acceptance criteria for USP <698> deliverable volume. The containers are labeled as 237 ml. Unfortunately, one of the initial ten samples was slightly less

Fig. 2.6 Minitab output for data in Table 3.2

Assay Category	Count	Percent	CumCnt	CumPct
0 - 44.5	3	2.50	3	2.50
45.5-49.4	23	19.17	26	21.67
49.5-50.5	57	47.50	83	69.17
50.6-55.5	31	25.83	114	95.00
55.6 +	6	5.00	120	100.00
N=	120			

Table 2.5 Results of samples for measuring deliverable volume

Initial sample (ml)	Second sample (ml)	
236.4	239.7	239.5
237.3	234.9	241.9
237.1	240.0	239.7
236.8	243.7	239.2
237.6	239.1	240.6
237.0	241.4	239.4
222.7	239.3	239.8
238.1	238.2	238.9
236.6	240.5	240.8
237.3	239.5	240.3

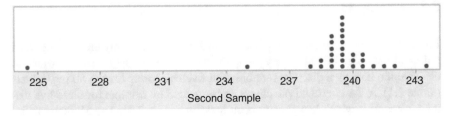

Variable	N	Mean	StDev	Minimum	Median
Initial Sample	10	237.69	4.59	224.70	239.05
Second Sample	30	239.11	3.07	224.70	239.45

Variable	N	SE Mean	CoefVar	Median	IQR
Initial Sample	10	1.45	1.93	239.05	0.83
Second Sample	30	0.561	1.28	239.45	1.17

Fig. 2.7 Minitab output for deliverable volume results in Table 2.5

than 95% of the labeled volume (less than 225 ml.). Therefore, an additional 20 containers are sampled and the results presented in Table 2.5. The second level criterial for 30 containers is that the average volume is not be less than the labeled amount and not more than 1 container be less than 95 percent of the labeled amount (USP Chapter <698>). Using Minitab (Fig. 2.7) the data shows that there is still only one bottle with less than 95% labeled amount (dot plot). The middle portion of Fig. 2.7 shows the selected values provided by Minitab and includes the mean, median, standard deviation, and the minimum value. Notice how close together the results are for the median and mean, indicating that the population is probably normally distributed. In this case the research would report the sample results as 239.11 ± 3.07 for all 30 samples. The lower portion of Fig. 2.7 provides additional information including the relative standard deviation (labeled *CoefVar*), the median, and interquartile range (in case there is a question of normality). In addition the *SE Mean* is reported. The importance of *SE Mean* (standard error of the mean) will be discussed in Sect. 3.5, but for the time being, notice the two measures are substan-

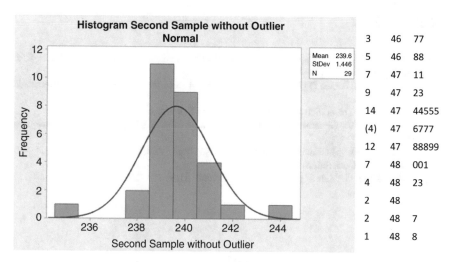

Fig. 2.8 Minitab histogram and stem-and-leaf plot of output for deliverable volume example (without the potential outlier 222.7)

tially different. What number is to be reported on the right side of the ± sign – *StDev* or *SE Mean*? For internal reporting purposes, the most common results would be to report the sample mean and standard deviation, but the standard error could be reported. In that case the research must be very careful in labeling the result to let the reader know it represents the standard error and not the standard deviation.

With Minitab, under "options" the researcher can select those descriptive statistics required (Table 2.4).

Visually a dot plot could be generated to observe the distribution of values. On the top of Fig. 2.7 is a plot with all 30 data points. As can be seen, there is only 1 failing value from the initial 10 samples and the final 30 (224.7 ml). This is an influential data point and possibly an outlier (Chap. 8). Creating a histogram for the other 29 data points without this potential outlier (Fig. 2.8) shows a much more symmetrical distribution. The results for a stem-and-leaf plot without the potential outlier is also presented on the right side of Fig. 2.8. With a mean volume greater than the labeled amount and only 1 out of 30 samples to fail, the results pass the criteria for USP <698>.

2.6   Inferential Univariate Statistics

There are several inferential tests that involve a single variable. These usually compare sample statistic (e.g., sample mean) to population parameters to determine if there is a difference. Examples of such tests include the one-sample t-test, sign test, and one-sample Z-test of proportions. These tests will be covered in Chap. 5, but before discussing the tests, it is important to understand the components of inferential tests (Chap. 3) and the importance of hypothesis testing (Chap. 4).

References

De Muth JE (2014) Basic statistics and pharmaceutical statistical applications, 3rd edn. CRC Press, Boca Raton, pp 114–116

Janson L, Fithian W, Hastie TJ (2015) Effective degrees of freedom: a flawed metaphor. Biometrika 102(2):479–485

Sturges HA (1926) The choice of a class interval. J Am Stat Assoc 21:65–66

USP 42-NF 37 (2019a) General chapter <698> deliverable volume. US Pharmacopeial Convention, Rockville

USP 42-NF 37 (2019b) General chapter <911> viscosity – capillary viscometer methods. US Pharmacopeial Convention, Rockville

Walker HM (1940) Degrees of freedom. J Educ Psychol 31(4):253–269

Chapter 3
Statistical Inference and Making Estimates of the Truth

Abstract To know how inferential tests work, it is important to understand the underpinnings associated with these tests. This chapter covers probability which is critical for all inferential tests and the likelihood of making a correct decision about a population based on a small sample taken from that population. This chapter considers characteristics associated with symmetrical and asymmetrical distributions. If it can be assumed that the population is symmetrically distributed, it is required that each level of the independent variable has similar dispersions. If the population is not normally distributed, nonparametric or distribution-free statistics should be considered. The center for the population can be estimated with a confidence interval, and tolerance limits can approximate a range of values representing the outcomes for the vast majority of the population.

Keywords Confidence interval · Homogeneity of variance · Nonparametric test · Normal distribution · Parametric test · Probability · Reliability coefficient · Standard error · Tolerance limits

Statistical inferences involve making statements about a population parameter based on the descriptive statistics for a sample taken from that population. To understand the inferential process, this chapter will briefly review probability, discuss the characteristics of the normal distribution, consider the role of a standard error measurement, and address how to deal with distributions that are not normally distributed. Inferential confidence intervals and tolerance limits will be discussed. Nonparametric tests will be introduced for dealing with ordinal data and non-normal situations.

© American Association of Pharmaceutical Scientists 2019
J. E. De Muth, *Practical Statistics for Pharmaceutical Analysis*, AAPS
Advances in the Pharmaceutical Sciences Series 40,
https://doi.org/10.1007/978-3-030-33989-0_3

3.1   Basic Elements of Probability

Probability can be considered the "essential thread" that runs throughout all statistical inference (Kachigan (1991), p. 59). Statements like "95% confidence" or "chance of error of less than 5%" are expressions of probability. The probability of an event $[p(E)]$ is the likelihood of that occurrence. It is associated with discrete variables. The probability of any event is the number of times or ways an event can occur (m) divided by the total number (N) of possible associated events (Fig. 3.1). A simple example would be the rolling of a single die from a pair of fair dice. There are six sides, each with a value, thus the probability of rolling a three is 1/6 or $p = 0.167$.

There are three general rules regarding all probabilities. First, a probability cannot be negative. Even an impossible outcome would have $p(E) = 0$. Second, the sum of probabilities of all mutually exclusive outcomes for a discrete variable is equal to one. For example, with the tossing of a coin, the probability of a head equals 0.50, the probability of a tail also equals 0.50, and the sum of both outcomes equals 1.0. Thus the probability of an outcome cannot be less than 0 or more than 1. The third general rule, because of the addition theorem, is that the likelihood of two or more mutually exclusive outcomes equals the sum of their individual probabilities (Fig. 3.1).

For any outcome E, there is a complementary event ($\bar{E}$), which can be considered "not E" or "E not." Since either E or $\bar{E}$ must occur, but both cannot occur at the same time, then $p(E) + p(\bar{E}) = 1$, written for the complement. This will become obvious when discussing hypothesis testing in the next chapter. If the desired confidence limit is 95% ($p = 0.95$), then the amount of acceptable error would have to be 5% ($p = 0.05$). Since the totaling probability is 100%, $p_{error} = 1.00 - 0.95 = 0.05$.

Probability of an event:	$$P(E) = \frac{m}{N}$$
Addition theorem per probability:	$$p(E_i \, or \, E_i) = p(E_i) + pE_j)$$
Sum of muturaly exclusive and exhaustive events:	$$\sum p(E) = 1.00$$
Compliment probability:	$$p(\bar{A}) = 1 - p(A)$$
Probability of two events Occurring (intercept):	$$p(A \cap B) = \frac{n(A \cap B)}{N}$$
Probability of either of two events occurring (conjoint)	$$p(A \cup B) = p(A) + p(B) - p(A \cap B)$$
Conditional probability:	$$p(B|A) = \frac{p(B \cap A)}{p(A)}$$

Fig. 3.1 Probability equations

Parts of Chap. 6 deal with relationships between discrete variables and will rely on *conditional probability*, where the probability of one event is calculated given that a certain level of a second discrete variable. For example, the probability can be calculated for event A occurring given the fact that only a certain level (or outcome) of a second variable (B) is considered (Fig. 3.1). This is used in the chi square test of independence (Sect. 6.5.2) where there should be no relationship between two discrete variables if the probability of event A is the same given event B_1 or B_2 or whatever number of B levels of the second variable and should be equal to the over all probability of event A. In other words, A is not influenced by B or A is independent of B.

For this book the proceed information about probability should be sufficient for inferential tests. There are many more rules for probability and expansions on the information provided here and the equations in Fig. 3.1. Potential sources for additional information are Daniel (2005), De Muth (2014), or Forthofer and Lee (1995).

3.2 Parametric vs. Nonparametric Inferential Statistics

The most commonly encountered inferential tests are termed parametric. Meaning they have requirements (parameters) in addition to the previously described criteria (Sect. 1.3), namely, that samples are representative of the population from which are taken and that each measurement is independent of any other data point. The two additional criteria are normality and homogeneity. Normality requires an expectation that the population from which the sample is taken is normally distributed. The second requirement is homogeneity of variance or *homoscedasticity*. Simply stated, the variances for each level of the discrete independent variable should be approximately equal.

Parametric procedures are very robust statistics and can handle slight deviations for the requirements of normality and homogeneity. The general rule is to use parametric procedures unless the sample is extremely skewed, bimodal, or rectangular or if there is very large differences in the sample variances. If either of the criteria are not met, an alternative nonparametric test should be considered. The last step for most of the tests in the flow charts in Appendix A asks about on normality and homogeneity (N/H?). If these requirements are not met, a nonparametric procedure should be used if available. In Chaps. 5 and 6, there nonparametric alternatives will be discussed following their parametric counterparts.

How does the researcher determine if a parametric procedure should be used and that normality and homogeneity exist?

3.3   Characteristics of a Normal Distribution

Described as a "bell-shaped" curve, the normal distribution represents a symmetrical spreading of data and is one of the most commonly occurring outcomes in nature, and its presence is assumed in several of the most commonly used statistical tests. Properties of the normal distribution have a very important role in the statistical theory of drawing inferences about population parameters (estimating confidence intervals) based on samples drawn from that population. The normal distribution is often referred to as the Gaussian distribution, after the mathematician Carl Friedrich Gauss, even though a formula to calculate a normal distribution was first reported by the French mathematician Abraham de Moivre in the mid-eighteenth century (Porter 1986).

There are certain characteristics that are common for all normal distribution (Fig. 3.2). First, the normal distribution is continuous, and the curve is symmetrical about the mean. Second, the mode, median, and mean are equal and represent the middle of the distribution. Third, since the mean and median are the same, the 50th percentile is at the mean with an equal amount of area under the curve, above and below the mean. Fourth, the probability of all possible outcomes is equal to 1.0; therefore, the total area under the curve is equal to 1.0. Since the mean is the 50th percentile, the area to left or right of the mean equals 0.5. Fifth, by definition, the area under the curve between one standard deviation above and one standard deviation below the mean contains an area equal to approximately 68% of the total area under the curve. At two standard deviations, this area is approximately 95%. Sixth, as distance from the mean (in the positive or negative direction) approaches infinity, the frequency of occurrences approaches zero. This last point illustrates the fact that most observations cluster around the center of the distribution, and very few data points occur at the extremes of the distribution. Also, if the curve is infinite in its bounds, it is not possible to set absolute external limits on the distribution.

The exact areas within the normal distribution can be measured using a normal standardized distribution or Z-distribution (Appendix A, Table 1). Using this table, it is possible to calculate the probability of being within any area under a normal distribution. The values are proportions of the curve (between that point and the mean) by expressing these proportions as percent of observations within a normal standardized distribution; there are certain important areas, measured as standard deviations (SD), that will be applied to confidence interval and other inferential statistics. These are:

$$90\% \text{ of data fall within} \pm 1.64 \text{ SD}$$
$$95\% \text{ of data fall within} \pm 1.96 \text{ SD}$$
$$99\% \text{ of data fall within} \pm 2.58 \text{ SD}$$

Fig. 3.2 Proportions between various standard deviations under a normal distribution

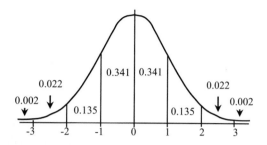

3.4 Determining Normality and Homogeneity

There are three ways to determine if the population is normally distributed. In all cases the sample distribution is the best and only estimation for the population distribution. A general rule is to assume a normal distribution unless the sample data shows a different spread (sample results are the best estimate of the population). The simplest method is to plot the data using a histogram or dot plot and see if the sample appears symmetrical. Remember that this is sample data and will not necessarily produce a perfect symmetrical curve. However, does it have a clustering of points near the center and a few data points to the extreme left and right of the center? If so, assume normality. A second method would be to examine the skew under Minitab's descriptive statistics (Table 2.4). The last method would be to perform a test of normality. Minitab offers applications to test for normality and these are discussed below.

3.4.1 Skew and Kurtosis

Skewness is a measure of asymmetry (Fig. 3.3). A zero value does not necessarily indicate symmetry (a bimodal distribution could produce a zero). A resulting negative value indicates a negatively skewed distribution, or tailing to the left. A positive value indicates a positively skewed distribution with tailing on the right side of the distribution (Fig. 2.4). The more positive or more negative the value, the greater the amount skewness. One common rule of thumb is that skew between -0.5 and $+0.5$ is approximately symmetrical; 0.5–1 is moderately skewed (positive or negative); greater than 1 is highly skewed (Blumer).

Kurtosis is a measure of the height of the peak of a distribution. A positive value indicates the relatively peeked distribution, with a sharper peak, shorter tails, and narrower shoulders. A negative value indicates a relatively flat distribution, with a shorter peak and longer tails. An average symmetrical distribution would be referred to a mesokurtic, a more peaked as leptokurtic, and a flatter distribution as platykurtic. Extreme kurtosis would be greater than $+3$ or less than -3.

Skew:

$$Skew = \frac{n}{(n-1)(n-2)} \sum \left[\frac{x_i - \bar{X}}{S} \right]^3$$

Kurtosis:

$$Kurtosis = \frac{n(n+1)}{(n-1)(n-2)(n-3)} \sum \left[\frac{x_i - \bar{X}}{S} \right]^4 - \frac{3(n-1)^2}{(n-2)(n-3)}$$

Fig. 3.3 Equations for calculating skew and kurtosis

With inferential statistics, there is less concern with kurtosis than skew. Regardless of kurtosis, if the population is estimated to be symmetrical, then parametric statistics can be applied.

3.4.2 Tests for Normality

There are a variety of ways to determine if the sample might be from a population with a normal distribution. The simplest would be to look at a histogram for the sample data. Does it look somewhat symmetrical? It is only a small subset of the population, but do the observations cluster around the center with a few data points to the extreme right and a few to the extreme left? If so, it can be assumed that the population probably has a similar pattern and is normally distributed.

A second method would be to look at the skew measurement. If the reported skew is greater than 1.0, caution should be used is assuming normality and an alternative nonparametric test might be preferred. Lastly there are several statistical tests for normality. Minitab includes three of these tests that will be discussed in Sect. 3.9.1.

3.4.3 Homogeneity of Variance

The second criteria for the use of a parametric inferential statistic is homogeneity. When comparing different levels of a discrete independent variable, the sample dispersions should be similar. For example, assume a comparison of two levels of an independent variable (i.e., control vs. experimental), and the variances were found to be 200 and 10, respectively. If representative of the population variance, they would certainly be different even if the centers might be identical. Therefore a parametric test would not be appropriate. Note that the concern is with variances, not standard deviations; so comparisons are made for the variances of the sample data. A quick rule of thumb for homogeneity of variance is that the largest variance should not be greater than twice the smallest variance. Minitab provides tests for the homogeneity of variances, and these will be discussed Sect. 3.9.2.

3.5 Distribution of Sample Means

Assume that there is a large population with numerous data points and samples are to be withdrawn at random from this population. Each sample would have descriptive statistics that include the sample mean and a sample standard deviation. These sample statistics would be the best estimates of the true population parameters.

$$\overline{X}_{sample} \approx \mu_{population}$$

$$S_{sample} \approx \sigma_{population}$$

The characteristics of dispersion or variability are not unique to samples alone. Individual sample means can also vary around the population mean. Just by chance, or luck, the samples could come from the upper or lower ends of the population distribution and calculated a sample mean that was too high or too low. Though there is no fault of the researcher, the estimate of the population mean would be erroneous.

Using the simple data in Table 1.1, assume that samples of 3 units are taken at random from the 20 observations. Using simple counting techniques, the total number of possible different samples taken three at a time would be 1140 samples (proof in Appendix D). The mean is a more efficient estimate of the center, because with repeated samples of the same size from a given population, the mean will show less variation than either the mode or the median. Statisticians have defined this outcome as the central limit theorem, and its derivation is beyond the scope of this book. However, there are three important characteristics that will be utilized in future statistical tests.

First, if the grand mean for all 1140 possible sample mean is calculated, it would equal the mean for the population of the 20 observations ($\mu_{\overline{x}} = \mu$). So, the center for all possible sample means is the center for the population. Second, the standard deviation for all 1140 possible sample means is equal to the population standard deviation divided by the square root of the sample size ($\sigma_{\overline{x}} = \sigma / \sqrt{n}$). This standard deviation for the means is referred to as the *standard error of the mean, SEM*, or simply the *standard error*. Finally, regardless of whether the population is normally distributed or skewed, if all the possible sample means are plotted, the frequency distribution will approximate that of a normal distribution, based on the *central limit theorem*. This theorem is critical to many statistical formulas because it justifies the assumption of normality. A sample size as small as $n = 30$ will often result in a near-normal sampling distribution (Kachigan (1991), p. 89). Using the example of data in Table 1.1, the grand mean for all 1140 samples would be 99.63% with a standard error of 0.42%.

In the reporting of descriptive statistics, Minitab includes *SE Mean* which is the standard error for the univariate sample. It is calculated by estimating σ using the sample standard deviation ($SE\,Mean = S / \sqrt{n}$). Notice in Fig. 2.7 that the standard error for the 30 samples (0.561) is considerably smaller than the standard deviation (3.07). Because the standard error is smaller, researchers might be tempted to use that value to indicate greater precision (Sect. 4.1). In reporting data, the mean and

standard deviation are usually the results that would be reported (239.11 ± 3.07 ml).
Be cautious of reports or articles that do not label the value to the left of the ± sign.
If not labeled, is the author reporting the standard deviation or the standard error?

As the sample size increases, the distribution becomes even more Gaussian. If it
approximates a normal distribution, based on the previous description of a normal
distribution, then approximately 68% of all sample means would be located between
one SEM above and below the mean. Also, other distributions for all the possible
sample means would include:

$$90\% \text{ fall within } \pm 1.64 \text{ SEM}$$
$$95\% \text{ fall within } \pm 1.96 \text{ SEM}$$
$$99\% \text{ fall within } \pm 2.58 \text{ SEM}$$

This knowledge of how sample means are distributed helps to establish the first
inferential statistic, a simple confidence interval.

3.6   Confidence Limits of the Mean

It is critical at this point to realize that this initial discussion is focused on the dispersion of data for the total population, not a sample. As mentioned in Chap. 2, in the
population, the mean is expressed as μ and standard deviation as σ. Sample data ($\bar{X}$
and S) are the best estimates of these population parameters, and the distribution of
the sample data (histogram or dot plot) provides the best estimator of the population
distribution. As noted in the previous section, the SEM and the central limit theorem
are based on a knowledge of the population standard deviation. Sometime the population variability is known, for example, during the production run of a pharmaceutical product that has been produced for many years. If the production protocol is
followed, the production engineer will know what standard deviation is expected to
be for a given run.

If the population dispersion is known, then any sample can be used to estimate
what the population mean might be using a confidence interval. All statistics are
either confidence interval or some type of a ratio. These two approaches will be seen
with the inferential tests in Chaps. 5 and 6. All confidence intervals use the same
formula:

$$\text{Population} = \frac{\text{Estimate}}{\text{(sample data)}} \pm \left(\begin{array}{c} \text{Reliability} \\ \text{Coefficient} \end{array} \times \begin{array}{c} \text{Standard} \\ \text{Error} \end{array} \right)$$

Various equations will follow this formula for tests in these two chapters. In this
case, where the population standard deviation is known or predicted, the population
mean can be estimated based on the best guess available (sample mean) plus or
minus the product of a reliability coefficient and a standard error term. The level of
confidence is based on an understanding of the normal distribution of all possible

means. The *reliability coefficient* is the amount of confidence the researcher requires. Based on the distributions presented in the previous section, if one wants to be 95% confident in the interval created, then 1.96 is used as the reliability coefficient (if 99% confidence, 2.58 is used). The standard error in this case is the estimate of the standard deviation for all possible sample means or the SEM ($\sigma / \sqrt{n}$). This specific confidence interval is referred to a *one-sample Z-test* and can be used when sigma (the population standard deviation is known or estimated) is presented in Fig. 3.4.

Assume in a hypothetical example that the results are $96.6 < \mu < 101.4$. The researcher would not be absolutely sure of the true population center, but would say with 95% confidence that the mean for the population is somewhere between 99.6 and 101.4 units of measure. An example using Minitab will be presented below.

Researchers may not know the population standard deviation. In this case an alternative method is requires and will be discussed in Chap. 5 as a one-sample t-test.

3.7   Tolerance Limits vs. Confidence Intervals

Confidence intervals estimate a range of possible values for the population mean (μ) based on sample data ($\bar{X}$). However, sometimes an investigator might be more interested in the approximate range of values for a particular population (e.g., tablets produced during a specific run). In most cases it is impossible to measure the entire population and know the "real-world" limits for all tablets produced during a specific run. However, we can determine limits within which we would expect to find 90%, 95%, or 99% of all the tablets produced. In this case, tolerance limits (or tolerance intervals) indicate the limits within which one would expect to find a given proportion of items from the population. Through statistical manipulation, with a certain degree of confidence, it is possible to calculate two values – the lower tolerance limit (LTL) and the upper tolerance limit (UTL). These limits represent the range where a given proportion of the population will exist. For example, during the production of a specific batch of tablets, with 95% confidence, what is the expected range of values (e.g., weight) for 99% of all the tablets in that batch? A sample is taken and the sample mean and standard deviation are calculated. Then formulas are used to estimate the LTL and UTL (Fig. 3.4). The result would be a range were 99% of the tablets would be expected to fall based on the attribute being measured; in this case tablet weights.

Confidence interval (where population standard deviation is known):	$\mu = \bar{X} \pm Z_{\alpha/2} \times \dfrac{\sigma}{\sqrt{n}}$
Tolerance limits: (upper and lower limits)	$LTL = \bar{X} - KS$
	$UTL = \bar{X} + KS$

Fig. 3.4 Equations for confidence intervals and tolerance limits

Tables for the reliability coefficients (*K*-values) can be found in Odeh and Owens (1980). Using these tables it is possible to determine where 95%, 99%, and 99.9% of the product would be with 90%, 95%, or 99% confidence. However, using Minitab to create the tolerance limit knowledge, these table values for the reliability coefficient are not needed.

Tolerance limits can be used to determine if a batch meets specifications (USP 42-NF 37 (2019), USP Chapter <1010>).

3.8 Nonparametric Alternative Tests

If there is concern about the normality or homogeneity, nonparametric alternative procedures may be considered. As noted preciously, following the flow chart in Appendix A for determining the most appropriate inferential statistic, in most cases the final questions is "N/H?". If yes to normality (N) and homoscedasticity (H), then use the parametric procedure. If not, the alternative nonparametric procedure is recommended.

Often referred to as *distribution-free statistics*, these tests can be useful when dealing with small sample sizes or when the requirement of a normally distributed population cannot be met or assumed. These tests are simple to calculate but are traditionally less powerful (statistical power will be discussed in Sect. 4.5). However, other authors (e.g., Conover 1999) argue that nonparametric tests are preferable and even more powerful than parametric tests if the assumptions (normality and homogeneity) are false. Nonparametric statistical tests do not make any assumptions about the population distribution. These tests are preferred when dealing with ordinal data associated with the dependent variable.

One of the limitations for nonparametric tests is the change from raw data to ranked information. With this change one loosed the fell and texture of the original continuous data and need to deal with relative positioning of the information.

3.8.1 Ranking Data Points

Most nonparametric tests involve ranking of the data from the smallest value (rank = 1) to the largest value (rank = N). For the statistical procedure, the ranks, rather than the original values, are used in the computations. A simple example of ranking is presented in Table 3.1. Notice identical values share the same average rank regardless of the condition where they are found (e.g., outcomes 94, 99, and 103). In the case of 94, since there are two identical values, they share the average of the ranks 4 and 5. Several nonparametric procedures are available with Minitab (Stats → Nonparametrics). If Minitab is used for any of these nonparametric tests, it will perform the ranking as part of the procedure.

Table 3.1 Example of ranking data for a nonparametric test

Experimental		Control	
Outcome	Rank	Outcome	Rank
89	1	93	3
90	2	94	4.5
94	4.5	96	7
95	6	98	8
99	10	99	10
100	12	99	10
101	13	102	14
103	15.5	103	15.5
106	17	109	19
107	18	112	20

3.9   Minitab Applications for Initial Inferential Statistics

Minitab can be used to determine whether a parametric or nonparametric test is most appropriate based on normality and homogeneity. The software can be used to create confidence intervals and determine tolerance limits.

3.9.1   Tests for Normality

As noted in Chap. 2, the simplest test for normality is to plot the sample data using a histogram or dot plot.

Graph → Histogram   or   Graph → Dotplot

Does the plot look skewed, multimodal (several high points), or rectangular? If so, it fails the test for normality. One of the options for a histogram in Minitab is the ability to "With Fit" that superimposes a bell-shaped curve over the histogram. This is presented in the example that follows.

The amount of skewness can help to determine the symmetry of the sample data by selecting "skew" as part of the descriptive statistics application.

Stats → Descriptive Statistics

For symmetrical distributions, the value should be less than or equal to 1.

Minitab offers three formal tests for normality: (1) Anderson-Darling, (2) Ryan-Joiner, and (3) Kolmogorov-Smirnov.

Procedure	Stats ➜ Descriptive Statistics ➜ Normality Test
Data input	Select continuous variables (columns in left box) from the worksheet.
Options	Select one of the three tests for normality. By default a scatter plot will be created.
Report	Any one of the three tests will create similar scatter plots with statistical information reported to the right of the graph. The statistical information will include the mean, standard deviation, sample size, test statistic requested, and associated p-value.
Interpretation	A p-value of less than 0.05 suggests that the results are not from a normally distributed population (see Sect. 4.3).

3.9.2   Tests for Homogeneity of Variance

There are several tests that measure the homogeneity of variance. Minitab uses Levene's test (Levene 1960) and a multiple comparison method which is a modification of Levene's test (Brown and Forsythe 1974). Levene's test is recommended for two discrete levels of the independent variable and the multiple comparison method for when there are more than two levels. These are located in Minitab under the ANOVA option.

Procedure	Stats ➜ ANOVA ➜ Test for Equal Variances
Data input	Select the source of information at the top of the dialog box: (1) data where each column is a variable on the worksheet ("Response" is the dependent variable and "Factors" is the independent variable); or (2) each level of the independent variable ("Responses") is in a different column in the worksheet.
Options	Automatic default is a 95% confidence interval. This can be changed to a different level of confidence. The assumption of a normal distribution can be added.
Graphs	The default is a summary plot, but individual value and box plots also are available.
Results	By default the method, confidence interval and test results will appear. Any of these can be removed using the "Results" option on the dialog box.
Storage	Sample means, standard deviations, and other options selected will appear in the next available columns on the worksheet.
Report	Results are presented based on the "Graphs" and "Results" selected.
Interpretation	For Levene's test or multiple comparisons, p-values less than 0.05 indicate that there is a lack of homogeneity.

3.9.3 Confidence Intervals Assuming a Known Population Standard Deviation

If the population standard deviation is known or can be estimated, a one-sample Z-test can be used to create a confidence interval. If the population standard deviation is unknown, an alternative procedure (1-Sample t) is presented in Sect. 5.2.1.

Procedure	Stat → Descriptive Statistics → 1-Sample Z
Data input	There are two ways to input data. The most common is data each variable in a separate column, and those columns are selected from the left box for evaluation. An alternative method would be for summary data where the sample mean and sample size are entered.
Known standard deviation	The population standard deviation must be known or can be estimated and entered in the space provided.
Hypothesized mean	Without a hypothesized mean, a confidence interval will be calculated. With a hypothesized mean, the program will calculate both a confidence interval and a Z-statistic with corresponding p-value.
Options	Automatic default for a 95% confidence interval for a two-tailed test (Sect. 5.1.1). These can be changed to a different confidence interval or a one-tailed test as desired.
Graphs	None by default, but histograms, individual value plot or box plot can be added.
Report	Without a hypothesized mean, a confidence interval is produced. With a hypothesized mean, a confidence interval will be created as well as a Z-statistic with corresponding p-value.
Interpretation	Without a hypothesized mean, a confidence interval is produced within which the population mean is expected with a given degree of confidence. With a hypothesized mean, reject the fact that the hypothesized mean falls outside the confidence interval or if the p-value is less than 0.05.

3.9.4 Tolerance Limits

Tolerance limits will compute a range of values for an attribute in the population based on sample results. Minitab has applications for estimated populations that are either normally distributed or not normally distributed. Assuming a normal distribution:

Procedure	Stat → Quality Tools → Tolerance Intervals (Normal Distribution)

Data input	There are two ways to input data. The most common is data each variable in a separate column, and those columns are selected from the left box for evaluation. An alternative method would be for summary data where the sample size, mean, and standard deviation are entered.
Minimum percent of population in interval	Automatic default for a 95% of total product but can change to another percent.
Options	Automatic default for a 95% confidence interval for a two-tailed test (Sect. 5.1.1). These can be changed to a different confidence interval or a one-tailed test as desired.
Graphs	For input from column data, the default is a summary plot with a histogram, parametric, and nonparametric tolerance limits, as well as a probability plot. These plots can be cancelled. For summary data, there are no graphics
Results	By default the method, confidence interval and test results will appear. Any of these can be removed using the "Results" option on the dialog box.
Storage	Information about the normal or nonparametric limits and/or tolerance factors can be stored in the next available columns on the worksheet.
Report	Tolerance limits for both parametric and nonparametric results. If data comes from a column, graphs are provided for the distribution, confidence interval, and probability plots.
Interpretation	Limits that are expressed represent product will fall (minimum percent of population in interval) with a certain degree of confidence (default of selected confidence limit under "Options") for both the normal and non-normal situation.

If a normal distribution in the population cannot be assumed, then the alternative selection is available (Tolerance Interval, Non-normal Distribution) where only data from a column is accepted. There are options for multiple types of distributions, but the easiest is to select "Assume none" and let the software select the best distribution. The other commands and output are similar to the normal distribution situation.

3.10 Examples

Twenty samples are taken at random during a scale-up run of an uncoated tablet. The results appear in Table 3.2. The Minitab results for the descriptive statistics are sample mean of 185.4 mg. with a standard deviation of 6.9 mg. Based on historical data for this product, the standard deviation for the population (σ) is expected to be 7.0.

Table 3.2 Weights (mg) of tablets from a scale-up run

187	192	182	184
184	186	178	190
174	187	185	179
193	176	179	182
183	201	188	197

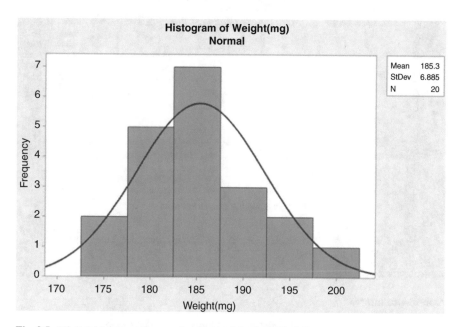

Fig. 3.5 Minitab histogram for sample tablet weights in Table 3.2

For normality, visually inspecting the histogram for this sample (Fig. 3.5), it appears to be a symmetrical distribution. A Minitab option has superimposed a symmetrical curve for this figure. The descriptive statistics option in Minitab gives a skew of 0.54 (Fig. 3.6) which is an indication of a slight skew. A further test for normality gives the results of Anderson-Darling test as $AD = 0.202$, $p = 0.859$ (greater than a p-value of 0.05) indication a symmetrical distribution. Thus, all three methods would indicate that the population is normally distributed based on the sample results.

Based on the sample, the researchers want to know what the average weight per tablet would be expected for the entire scale-up (population mean). A one-sample Z-test would be performed assuming that previous tests have estimated the population standard deviation to be 7.0 mg. The result would be $182.3 < \mu < 188.4$ mg (Fig. 3.6). The researchers would present the reportable

Descriptive statistics:

Variable	N	Mean	StDev	Sum	Median	Skewness
Weight(mg)	20	185.35	6.88	3707.00	184.50	0.54

Test for normality:

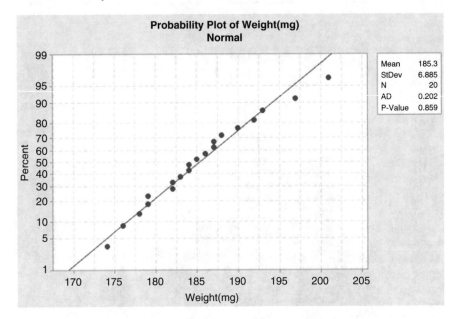

Confidence interval:

N	Mean	StDev	SE Mean	95% CI for μ
20	185.35	6.88	1.57	(182.28, 188.42)

μ: mean of Weight(mg)
Known standard deviation = 7

Tolerance limits:

Variable	Normal Method	Nonparametric Method	Achieved Confidence
Weight(mg)	(166.346, 204.354)	(174.000, 201.000)	26.4%

Achieved confidence level applies only to nonparametric method.

Fig. 3.6 Minitab results evaluating randomly sampled tablets in Table 3.2

Variable	N	Mean	StDev	Variance
Sample 1	20	185.35	6.88	47.40
Sample 2	20	187.40	7.95	63.23

Test for homogeneity:

Method	Test Statistic	P-Value
Multiple comparisons	0.27	0.605
Levene	0.33	0.568

Fig. 3.7 Minitab results homogeneity of variance between two groups

values from the sample to be 185.4 ± 6.9 mg. and with 95% confidence predict that the average weight per tablet for the entire scale-up run to be somewhere between 182.3 and 188.4 mg.

For the tolerance limit (also in Fig. 3.6), the expected range of weights for 95% of the tablets produced, with 95% confidence, would be between 166.35 and 204.35 mg.

As a second example, assume that an additional sample of 20 tablets are collected and intended to be compared to the previous example for sample weights. The researcher wanted to know if there was homogeneity of variance before deciding on a parametric or nonparametric test to compare these two samples. As seen in Fig. 3.7, second sample had a sample mean of 187.40 mg. and standard deviation of 7.95. In this case the two variances are 47.40 and 62.23 (the standard deviations squared). Using the simple rule of thumb would indicate that there is homogeneity (47.40/63.23 is less than 2). Levene's test in Fig. 3.7 confirms this decision ($p = 0.568$). Because the p-value is greater than 0.05, there is failure to reject homogeneity. The importance of the p-value will be discussed in the next chapter.

References

Brown MB, Forsythe AB (1974) Robust tests for the equality of variances. J Am Stat Assoc 69(364):364–367

Conover WJ (1999) Practical nonparametric statistics. John Wiley and Sons, Inc., New York, p 80

Daniel WW (2005) Biostatistics: a foundation for analysis in the health sciences, 7th edn. John Wiley and Sons, New York, pp 59–85

De Muth JE (2014) Basic statistics and pharmaceutical statistical applications, 3rd edn. CRC Press, Boca Raton, FL, pp 19–33

Forthofer RN, Lee ES (1995) Introduction to biostatistics. Academic Press, San Diego, pp 93–101

Kachigan SA (1991) Multivariate statistical analysis, 2nd edn. Radius Press, New York, p 59

Levene H (1960) Robust tests for equality of variances. In: Olkin I, Harold Hotelling et al (eds) Contributions to probability and statistics: essays in honor of Harold Hotelling. Stanford University Press, Palo Alto, pp 278–292

Odeh RE, Owen DB (1980) Tables for normal tolerance limits, sampling plans, and screening. Marcel Dekker, Inc., New York, pp 90–93, 98–105

Porter TM (1986) The rise of statistical thinking. Princeton University Press, Princeton, NJ, p 93

USP 42-NF 37 (2019) General Chapter <1010> analytical data – interpretation and treatment. US Pharmacopeial Convention, Rockville, MD

Chapter 4
Dealing with Inherent Statistical Error

Abstract When performing an inferential statistical test, one can never be absolutely sure that the results are correct. Through careful study design and with good scientific technique, systematic error can be controlled, but random error will always be present. This chapter focuses on hypothesis testing and types of random error associated with incorrect decisions: (1) rejecting a true hypothesis under test or (2) failing to reject a false hypothesis under test. The goal is to reject the hypothesis under test with at least 95% confidence in the decision or less than a 5% chance of being wrong. Another goal is to reject the hypothesis under test when that hypothesis is false, which is a called statistical power. Minitab applications to determine sample size and power are discussed.

Keywords Accuracy · Alternative hypothesis · Hypothesis testing · Null hypothesis · *p*-value · Precision · Random error · Sample size · Statistical power · Systematic error

Inferential statistical tests involve making an estimate about a population based on data collected in a sample. Through well-designed studies, we can remove most of the potential error, but we can never be 100% sure of the results of such a statistical analysis. This chapter deals with potential errors and how to minimize their effects and includes a focus on random and systematic errors, hypothesis testing, statistical power, and sample size determination.

4.1 Random vs. Systematic Error

Sample data should be representative of the true population which is why good sampling plans are so important. *Precision* refers to how closely data are grouped together or the compactness of the sample data. Figure 4.1 presents data with different amounts of scatter. Samples A and C have less scatter or more closely clustered

© American Association of Pharmaceutical Scientists 2019
J. E. De Muth, *Practical Statistics for Pharmaceutical Analysis*, AAPS
Advances in the Pharmaceutical Sciences Series 40,
https://doi.org/10.1007/978-3-030-33989-0_4

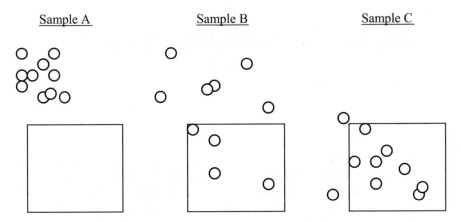

Fig. 4.1 Samples comparing precision and accuracy

data and represent greater precision. Sample B has a great deal of scatter and represents less precision. Precision measures the variability within a group of measurements, and greater precision would be represented by smaller standard deviations or smaller relative standard deviations.

However, assume that the smaller boxes in Fig. 4.1 represent the true value for the population from which the samples were taken. In this example, even though Samples A and C have good precision, only Sample C is an accurate reflection of the population. *Accuracy* is concerned with "correctness" of the results and refers to how closely the sample data represents the true value of the population. It is desirable to have data that is both accurate and precise. Sample A is probably the most precise, but it lacks accuracy. This lack of accuracy is called bias.

Ideally, sample data should have both good accurate and good precise. The question is how to assess these two components?

Bias can be thought of as some type of *systematic error*. It is something that causes a constant error in the measurement. In the example of a pharmaceutical analysis, it might be due to a calibration error or an error in the equipment operator's technique that would cause a constant error. At the same time, there could be a *selection bias* where certain characteristics make potential data more (or less) likely to be included in the study. Ideally, investigators should use random sampling or a specific probabilistic sampling plan to avoid selection bias. For example, always sampling from the top of storage drums may bias the results based on particle size, assuming smaller particles settle to the lower regions of the drums. Control of systematic error is the responsibility of the analytical scientist, not the statistician, as part of a good study design and research protocol. The determination of systematic error and lack of accuracy would be to compare the sample results to a reference standard or a well-characterized procedure (FDA, 1996). In this case a simple confidence interval (discussed in Sect. 3.6) could be used to compare sample results with a reference standard target. Does this target fall within the 95% confidence interval? Here we need an estimate of the population standard deviation (σ) based on history to create the confidence interval. If this information about the population variability is not available, a one-sample t-test can be used with the sample standard deviation (Sect. 5.2). The accuracy of the test is the closeness of the results to the reference standard target.

As mentioned previously, precision is determined by the size of the standard deviation or relative standard deviation. The smaller these values, the greater the precision. For example, many of the tests for content uniformity in USP Chapter <905> require results to be between certain percent (e.g., 85–105%) of label claim and have an RSD equal to or less than 6% (USP 41-NF 36 (2018a), <905>). In this case the percent is a measure of required accuracy, and the RSD is the required precision. The type of error associated with precision is called *random error*. For analytical data, precision is the amount of agreement seen with repeated tests on a homogeneous sample. When researchers are comparing an alternative method to the current analytical method, the two methods can be considered comparable if the precision of the new method is not be worse than that of the current method by an amount deemed important. In this case "worse" would be defined as increased variability (decreased precision). How much of a difference is deemed important? Sample variances (S^2) are usually the only measurements we have for the precision of two methods. A general rule of thumb is differences in precision are important if there is twofold difference between the standard deviations for the samples. Since the variance is the square of the standard deviation, a magnitude (M) of a fourfold difference (S^2) is considered important. Similarly, USP Chapter <1010> provides a method for comparison of precisions where the desired M is less than four (USP 41-NF 36 (2018b), <1010>). The equations for calculating precision using the USP method are presented in Fig. 4.2, and the upper limit would need to be less than four to have acceptable precision for the alternative method. For example, assume we have two methods, where the new method ($S^2 = 3.151$, $n = 12$) and the existing method ($S^2 = 4.215$, $n = 12$) are being compared. The result would be $0.265 < M < 2.106$. In this case the M-value is less than four, and the conclusion would be that similar precision exists for both methods.

Precision and accuracy are two important considerations when dealing with method validation. The objective of validating an analytical procedure is to demonstrate that it is suitable for its intended purpose (ICH 2005, p.1). The criteria for a validated method not only include accuracy and precision but also include specificity, detection limits, quantitation limits, linearly, range, and robustness (USP <1225>). Specificity is simply the ability of an assay to measure that particular substance it is intended to measure and not any other substances (e.g., excipients, impurities, degradants) that might be in the sample. It is usually a pass-fail criterion and no statistics are involved. The limit of detection (LOD) is the lowest amount of analyte in a sample which can be detected but not necessarily quantitated as an exact value. A signal-to-noise ratio between 3:1 and 2:1 is acceptable for estimating the detection limit (ICH 2005, p. 11). It can be calculated if the population standard deviation is known, and there is a known calibration curve (m), LOD = $3.3\sigma/m$. The limit of quantitation (LOQ) is the lowest amount of analyte in a sample which can be quantitatively determined with suitable precision and accuracy. A signal-to-noise

$$S_{ratio}^2 = \frac{S_{New\ Method}^2}{S_{Established\ Method}^2} \approx \frac{\sigma_{New}^2}{\sigma_{Established}^2}$$

$$Lower\ Limit\ (M_L) = \frac{S_{ratio}^2}{F_{0.95,n_1-1,n_2-1}}$$

$$Upper\ Limit\ (M_U) = \frac{S_{ratio}^2}{F_{0.05,n_1-1,n_2-1}}$$

Where $F_{.05}$ and $F_{.95}$ are values taken from the table of critical F-values presented in Table B3, Appendix B. The table does not provide values for $F_{0.05}$ but can be calculated as $F_{0.05,n_1-1,n_2-1} = 1/F_{0.95,n_2-1,n_1-1}$.

Fig. 4.2 Equations for comparing precision of a sample to a reference

ratio between 10:1 is acceptable for estimating the quantitation limit (ICH 2005, p. 12). Again, with known parameters, it can be calculated as LOQ = 10σ/m.

Linearity will be discussed in Chap. 6, but the range (introduced in Chap. 2) in this case represents the upper and lower concentrations of analyte with demonstrated suitable levels of precision, accuracy, and linearity. So all three must be tested within this range; ICH Q2 recommends the following minimum ranges for a drug substance or drug product of 80–120%: for content uniformity to be 70–130%; for dissolution testing at ±20% over the specified range; and for impurities 1–120% of the specification (ICH 2005, p. 8).

Lastly robustness is the ability of an analytical procedure to remain unaffected by small but deliberate variations in test conditions and indicates a certain reliability during normal usage. Small variations could include assay time, pH, temperature, sample preparation, or different lots of reagents. Statistical tests described in Chap. 5 could be used to determine if these small deliberate changes cause significantly different results.

Various test methods are used for assessing compliance of pharmaceutical articles, and these should have established specifications and must meet proper standards of accuracy and reliability. *Reliability* is a collection of factors and judgments that, when taken together, are a measure of reproducibility. It is the consistency of measures and deals with the amount of error associated with the measured values. In order for data to be reliable, all sources of error and their magnitude should be known, including both constant errors (bias) and random (chance) errors. *Validity* refers to the fact that the data represents a true measurement. A valid piece of data describes or measures what it is supposed to represent. It is possible for a sample to be reliable without being valid, but it cannot be valid without being reliable. Therefore, the degree of validity for a set of measurements is limited by its degree of reliability. Also, if randomness is removed from the sampling technique used to collect data, it potentially removes the validity of our estimation of a population parameter.

4.2 Hypothesis Testing

Hypothesis testing is the process of inferring from a sample whether to reject a certain statement about a population or populations. The sample is assumed to be a small representative proportion of the total population and based on a good sampling plan involving a protocol that removes as much systematic error as possible. Hypothesis testing deals solely with random error. Hypotheses are established, and two possible errors can occur: (1) rejection of a true hypothesis or (2) failing to reject a false hypothesis.

Sometimes referred to as *significance testing*, hypothesis testing involves two statements:

> Hypothesis: Fact A
>
> Alternative: Fact A is false

Researchers must carefully define the population about which they plan to make an inference. For example, if 30 capsules were drawn at random from one specific batch of a product and some analytical procedure was performed on the sample, this measurement could be considered indicative of only one population (the batch) and cannot be generalized to other batches of the same medication. In addition, with any inferential statistical test, it is assumed that the individual measurements are independent of one another and any one measurement will not influence the outcome of any other member of the sample. Also, the stated hypotheses should be free from apparent prejudgment or bias. Lastly, the hypotheses should be well-defined and clearly stated. Thus, the results of the statistical test will determine which hypothesis is determined to be correct. All of these requirements are related to good science and careful data collection on the part of the researcher and not the responsibility of the statistician.

The initial hypothesis may be rejected, meaning the evidence from the sample casts enough doubt about the hypothesis that it can be said with some degree of certainty that the hypothesis is false. If this initial hypothesis (called the *null hypothesis*) is rejected, an *alternative hypothesis* is required. This alternative hypothesis is the statement the researcher is usually trying to prove.

> H_0: Null hypothesis (*hypothesis under test*)
>
> H_1: Alternative hypothesis (*research hypothesis*)

By convention, the null hypothesis is stated as no real differences in the outcomes or a relationship of zero (a null relationship). Examples of null hypotheses that will be discussed in later chapters are presented in Fig. 4.3. The alternative hypothesis would be statements that are mutually exclusive and exhaustive for the null hypothesis. For example, if we are comparing two levels of a discrete independent variable ($\overline{X}_1$, $\overline{X}_2$), the null hypothesis would be stated as $\mu_1 = \mu_2$. Notice that the available data are sample means, but the actual hypothesis being tested involves inference about the population means. The evaluation then attempts to nullify the hypothesis of no significant difference in favor of an alternative research hypothesis ($\mu_1 \neq \mu_2$).

Chapters	Statistical Tests	Null Hypothesis
5	Two-sample t-test	$\mu_1 = \mu_2$ or $\mu_1 - \mu_2 = 0$
5	Paired t-test	$\mu_d = 0$
5	Analysis of variance	$\mu_1 = \mu_2 = \mu_3 = \ldots \mu_k$
6	Correlation	$r_{xy} = 0$
6	Linear regression	No linear regression
6	Tests of association	No association
5 and 6	Nonparametric tests	Same population

Fig. 4.3 Examples of null hypotheses

The type of null hypothesis will depend upon the types of variables and the outcomes the researcher is interested in measuring. In Chaps. 5 and 6, there will be presented different hypothesis based on the type of inferential statistic being used.

As noted, the two hypotheses must be mutually exclusive and exhaustive. They cannot both occur, and there is no third possible result.

H_0: Hypothesis A
H_1: Hypothesis A is false

Statistics from the sample data (sample means and standard deviations) provide the information used to estimate the probability that some observed difference (Chap. 5) or relationship (Chap. 6) between samples should be expected due to sampling error. There are two approaches to inferential statistics (1) create a confidence interval or (2) establish a ratio and compare the resultant test statistic to a predetermined critical value. These are discussed in the next section.

4.2.1 Evaluating Significance

The first approach for deterring significance has already been employed in the previous chapter (Sect. 3.6), with the establishment of a confidence interval to estimate a population parameter based on sample results.

$$\text{Population} = \frac{\text{Estimate}}{(\text{sample data})} \pm \left(\begin{array}{c} \text{Reliability} \\ \text{Coefficient} \end{array} \times \begin{array}{c} \text{Standard} \\ \text{Error} \end{array} \right)$$

The result is an interval, and the rejection of the null hypothesis is based on whether or not the desired result is within the interval. In the previous chapter, the result of the confidence interval was a range of possible values for the true population mean. In hypothesis testing a simple null hypothesis might be $H_0 : \mu_1 = \mu_2$. As seen in the next chapter (Sect. 5.2.1), an identical hypothesis would be $H_0 : \mu_1 - \mu_2 = 0$. For this particular test, a confidence interval is used to evaluate this hypothesis, and the calculated confidence would be expected to include zero (the null hypothesis).

However, if the alternative hypothesis is true ($H_1 : \mu_1 - \mu_2 \neq 0$), the zero will not fall within the confidence interval. Whether or not the interval includes zero will determine which hypothesis is accepted.

The second method calculates a "test statistic" (a value based on the manipulation of sample data). This value is compared to a preset "critical" value (usually found in a special table) based on a specific acceptable error rate (e.g., 5% or $p = 0.05$). In most cases this involves a ratio, simplified to the following:

$$\text{Test statistic} = \frac{\text{Estimate}\left(\text{sample data}\right)}{\text{Standard Error}}$$

If the outcome is extremely rare, it will be to the extreme of our critical value, and we will reject the hypothesis under test in favor of the research hypothesis. As will be seen in the following chapters, almost all inferential statistics use one or both of these approached.

4.2.2   Rejecting or Failing to Reject the Null Hypothesis

The statistical test results have only two possible outcomes, either the null hypothesis cannot reject or it can be rejected in favor of the alternative hypothesis. At the same time, if all the facts were known (the real world) or we had data for the entire population(s), the null hypothesis (H_0) is either true or false for the population(s) that the sample represents. This is represented at the top of Fig. 4.4. However, along the left margin are two possible outcomes from a statistical test. The goal of inferential statistics is that results will fall into the upper left and lower right quadrants of this two-by-two table. Results in the lower left and upper right areas are deemed errors.

There is analogy to hypothesis testing that can be seen in American jurisprudence (De Muth 2014). Where the hypotheses are H_0, person is innocent of the crime, and H_1, person is guilty of the crime. During the trial, the jury will be presented with a sample of the truth (information, exhibits, testimonies, evidence) that will help, or hinder, their decision-making process. Hopefully the jury will reject the null hypothesis if the person is truly guilty and find the person not guilty if truly innocent (avoiding the two types of errors in Fig. 4.4). Note that in this analogy, if the jury fails to find the person guilty, their decision is not that the person is "innocent." Instead they render a verdict of "not guilty" (they failed to have enough evidence to prove guilt). In a similar vein, the decision is not to accept a null hypothesis, but to *fail* to reject it. If we cannot reject the null hypothesis, it does not prove that the statement is actually true. It only indicates that there is insufficient evidence to justify rejection of the null hypothesis.

As illustrated in Fig. 4.4, there are two possible random errors associated with hypothesis testing. *Type I error* is the probability of rejecting a true null hypothesis, and type II error is the probability of accepting a false null hypothesis. Type I error is also called the *level of significance* and uses the symbol α or p. Why two symbols?

The Real World (Population)

		H_0 is true	H_0 is false
Statistical Test (based on sample data)	Fail to Reject H_0	$1-\alpha$ Confidence	β Type II error
	Reject H_0	α, p Type I error	$1-\beta$ Power

Fig. 4.4 Illustration of possible outcomes from hypothesis testing

One symbol (α) is an *a prior* decision before the test is run. The researchers may want to be 95% confident of the results if the decision is to reject the null hypothesis; here $\alpha = 0.05$ and established before the study. But after the statistical program is completed, most computer programs will report a test statistic value and an associated p-value, which represents the amount of type I error after the fact. This value is the true amount of error the researcher has to be willing to accept to reject the null hypothesis. For example, assume an analyst is comparing two methods and he want to be 95% confident in his decision to reject the null hypothesis that the two methods are identical, if a difference exists. In this case $\alpha = 0.05$. However, upon completion of the test, Minitab software reports $p = 0.003$. Here the researcher can say with 99.7% confidence that the two methods are different. His chance of being wrong is only 0.3%, well less that the a priori type I error rate of 5%.

Similar to the left two boxes in Fig. 4.4, there are two complimentary boxes on the right side that also total 100% ($p = 1.00$). *Type II error* is symbolized using the Greek letter beta (β) or a failure to reject the null hypothesis when it is false. The probability of rejecting a false H_0 is called *power* ($1 - \beta$). In hypothesis testing the goal is to minimize the type I error and maximize the statistical power. The most common convention is an acceptable type I error is 5% or less and adequate power would be 80% or greater.

4.2.3 Minitab Reporting of Type I Error

Minitab is an excellent program, but in the author's opinion, there is one major error with the way type I error (the p-value) is reported. For whatever reason, with extremely small p-values, Minitab truncates the results to report $p = 0.000$. Which is impossible; one can never be 100% confident (zero percent chance of being wrong) in the decision to reject the null hypothesis. For example, the one-way

ANOVA example presented in Sect. 5.5.6, the results on the Minitab printout (Fig. 5.22) are $F = 8.89$ and $p = 0.000$. However, evaluating the exact same data using Excel, the results are $F = 8.89$ and $p = 0.0000037$. Minitab's reporting $p = 0.000$ is incorrect. The user of Minitab should not report the value $p = 0.000$ but instead $p < 0.001$; because with the Minitab results, it is impossible to determine what exists past the third zero. This approach for reporting small p-values will be used in the next three chapters when interpreting the Minitab reported results.

4.3 Power and Sample Size Determination

Type II error and power are closely associated with sample size and the amount of difference the researcher wishes to detect. Power is the complement of type II error (β); decrease the type II error and the statistical power of the test will increase. Power is more difficult to control than type I error, where one simply selects a value from a statistical table for the amount of error one is willing to tolerate in rejecting a true null hypotheses. Type II error and power are affected by (1) sample size; (2) size of the detectable difference of interest; (3) amount of dispersion (standard deviation or variance); and (4) even *a prior* type I error selected for the analysis. All are interrelated:

$$\text{Type II Error} = \frac{\text{Detectable difference}}{\sqrt{\dfrac{\text{Dispersion}}{\text{Sample Size}}}} + \text{Type I Error}$$

To illustrate the importance of each one of these factors, Figs. 4.5, 4.6, 4.7, and 4.8 show what happens to statistical power when each factor is changed, keeping the remaining three factors constant.

One way to reduce both types of error is to increase the sample size. Larger sample sizes will result in greater power and less type II error (Fig. 4.5). Small sample sizes generally lack statistical power and are more likely to fail to identify important differences because the test results will be statistically insignificant. This makes sense; the more one knows about the population (larger sample size), the more confidence in the decision of the inferential test.

Obviously, if a difference exists, it is easier to detect large differences than very small ones. As seen in Fig. 4.6 when trying to identify larger differences, power will increase. The difficulty here is defining what amount of difference is important. In the clinical environment, it might be easier to define as a "clinically significant difference." In the analytical arena, this is more difficult. For example, if comparing the results between two laboratories doing the same test, how large a difference between the two laboratories would be significant? Should it be two percent, five percent, ten percent, and twenty percent? Here is another example of a question that must be addressed by the scientist, not the statistician.

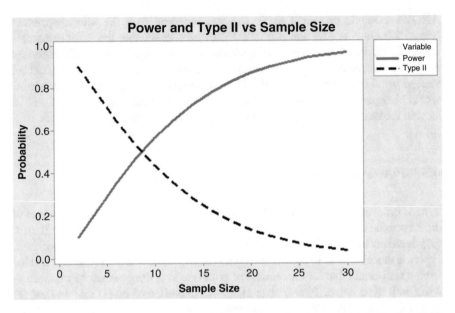

Fig. 4.5 Effect of changes in sample size on statistical power (constants of $\delta = 6.5$, $\sigma^2 = 42.25$, $\alpha = 0.05$)

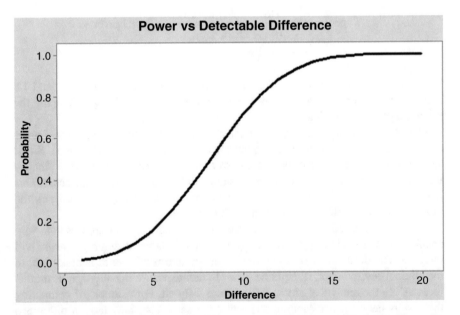

Fig. 4.6 Effect of changes in detectable differences on statistical power (constants of $n = 10$, $\sigma^2 = 42.25$, $\alpha = 0.05$)

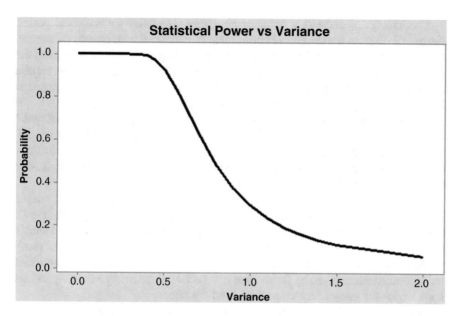

Fig. 4.7 Effect of changes in variance on statistical power (constants of $n = 10$, $\delta = 1.0$, $\alpha = 0.05$)

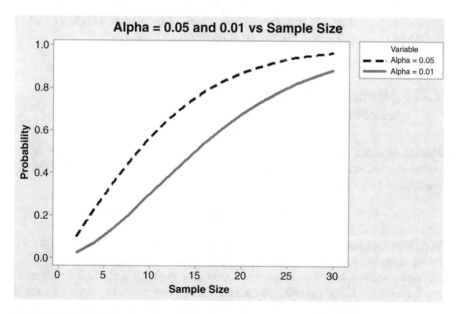

Fig. 4.8 Effect of changes in sample sizes on statistical power for two levels of type I error (constants of $\delta = 6.5$, $\sigma^2 = 42.25$)

The amount of dispersion or uncertainty can also influence statistical power. Figure 4.7 displays the decrease in power that is associated with greater variance in the sample data. Conversely, as the dispersion within the population decreases, the power of the test will increase. As will be seen, the problem is most formulas require a known population standard deviation (σ). Which in most cases will either need to be estimated or approximated using the sample standard deviation (S).

Figure 4.8 illustrates changes in power for two different a priori levels of type I error with increasing sample sizes. As we increase our confidence that there is a difference (making α smaller), there is an increase of chance of missing a true difference, increasing β or decreasing power. So changing a type I error of 0.05 to 0.01 or 0.005 will actually decrease statistical power.

Numerous formulas exist that can be used to calculate the appropriate sample size under different criteria. These include power curves presented by Kirk (1968), based on α, $1 - \beta$, and the number of levels of the independent variable, and Young's nomograms (1983) for sample size determination. An excellent reference for many of these methods is presented by Zar (2010). A discussion of power and sample size for binomial tests is given by Bolton and Bon (2004). Fortunately Minitab provides applications for many of the commonly used statistics.

With 95% confidence we can reject a null hypothesis, but as mentioned earlier we never prove the null hypothesis. This creates a problem especially if we would like to prove that one level of an independent variable is similar (or not statically different) to another level. One can never prove they are exactly the same, but it is possible to prove they are similar with 5% or 10% of each other. This topic will be addressed in Chap. 7 with a discussion of equivalence tests.

4.3.1   Minitab Applications for Determining Sample Size and Power

Minitab provides sample size determinations and power calculation for many of the commonly encountered inferential tests in Chap. 5: one-sample t-test, two-sample t-test, paired t-test, and one-way ANOVA.

Stats → Power and Sample Size → *List of tests available*

Most tests will have a common initial dialog box that is illustrated in Fig. 4.9 for a two-sample t-test. There are three open boxes at the top: sample size, differences, and power values. Any two of these areas must be completed, and Minitab will solve for the third field. For example, to estimate power, one needs to provide a sample size and detectable differences. The resultant output will be the power expressed as a proportion.

Multiple results can be tested at one time. For example, assume the researcher wants to try different sample sizes to determine what size is best with respect to statistical power. The researcher would enter the detectable difference and several different sample sizes each separated by a space (e.g., 5 10 15 20 25 30). The resul-

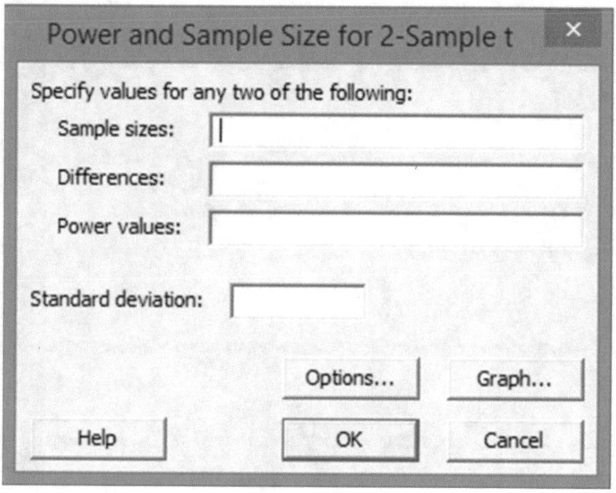

Fig. 4.9 Example of a Minitab main dialog box for power and sample size determination

tant output would show columns with the detectable difference, sample sizes, and corresponding power for each sample size.

The last box at the bottom of the main dialog box is "standard deviation." In distant earlier versions of the software, it was labeled as "sigma" which is the correct term (sample size determination requires an estimate of the population standard deviation). This box must have a value, or the program will not run; so some estimate of dispersion is critical. If no prior information is available about an expected population standard deviation, the best estimate is the sample standard deviation. If comparing one material to a gold standard and the population standard deviation is unknown, then standard deviation for the gold standard is used. If comparing two or more discrete groups, then the average for all the standard deviations is used in the lower box.

For power and sample size determination, Minitab set the default type I error at 0.05. This can be overridden by selecting "Options" in most tests and changing the "Significance level" to something other than the default of 0.05. By default the determination will be based on a two-tailed distribution (Sect. 5.1.1); this can be changes under "Options." Also by default, Minitab will provide a graphic of various power curves, as illustrated in Fig. 4.10.

4.3.2 Examples

In the previous chapter (Sect. 3.10), a simple Z-test was used to estimate the population weights of tablets based on an initial sample of 20 tablets. The result was a 95% confidence interval that included population weights were between 182.3 and 188.4 mg. Assume that the researcher wished to be able to detect a 2% difference from the desired weight of 186 mg (2% would be 3.72 mg). As noted, historically

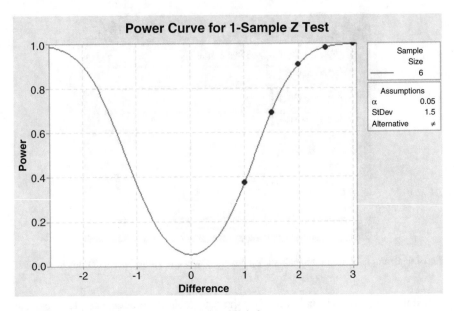

Fig. 4.10 Example of a power curve created by Minitab

the dispersion on such a batch would be $\sigma = 7$ mg. How many samples should have been taken to have power of 80%? For this example, different sample sizes are entered in to Minitab (n = 10, 15, 20, 25, 30, 35, and 40). The results for a power determination for a one-sample Z-test are presented in Fig. 4.11. Unfortunately, a sample size of 20 units had the power of only 0.662. As the sample size increased, power increases, and if the researcher wants power to be ≥ 0.80, a sample size of 30 was needed. The power curves for the various sample sizes are presented in Fig. 4.12.

For a second example, assume that the protocol calls for a sample size of six, how large a difference could be detected with power of 80%. In this case, values of 1, 1.5, 2, 2.5, and 3 would be entered under "differences," 0.80 for power, 6 for sample size, and 1.5 for standard deviation (based on previous experience with the product). The results are seen in Fig. 4.13. Here with a sample size of six, the smallest detectable difference with the required power would be somewhere between 1.5% and 2.0%. The representative power curves for these results has already been presented in Fig. 4.10.

4.4 Propagation of Error

One should always consider the potential compounding of error during an experiment. As sum that an analytical method or piece of equipment has an inherent about of error or variability. Sample data is collected, and this sample also has variability, as measured by the standard deviation. The results are a large amount of

Fig. 4.11 Minitab output of statistical power for different detectable differences

1-Sample Z Test
Testing mean = null (versus ≠ null)
Calculating power for mean = null + difference
α = 0.05   Assumed standard deviation = 7

Difference	Sample Size	Power
3.72	10	0.390090
3.72	15	0.539162
3.72	20	0.661543
3.72	25	0.757157
3.72	30	0.829145
3.72	35	0.881796
3.72	40	0.919406

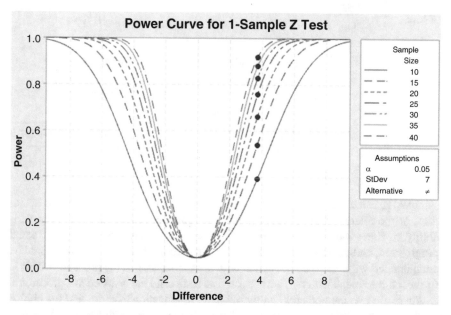

Fig. 4.12 Minitab power curves for various sample sizes

total variability than just that seen in the sample. Other times the final results for an experiment are not measurable, but are the results of some adding, subtracting, multiplying, or dividing of the results of the other original measurements. It becomes necessary to estimate the total amount of error based on these types of mathematical manipulations. This combining of uncertainties from separate measures is referred to as *propagation of error*. It is the resultant measure of dispersion where the results are dependent on a number of different independent variables, each of which is measured. Each independent variable will be associated with the total measure of uncertainty (error). For example of with surface areas, the error components are estimated from repeating the measurement several times (or taking numerous

Fig. 4.13 Minitab output of statistical power for different detectable differences

1-Sample Z Test
Testing mean = null (versus ≠ null)
Calculating power for mean = null + difference
α = 0.05   Assumed standard deviation = 1.5

| | Sample | |
Difference	Size	Power
1.0	6	0.372008
1.5	6	0.687770
2.0	6	0.904228
2.5	6	0.983103
3.0	6	0.998354

Fig. 4.14 Equations for calculating the propagation of error

Serial propagation involving addition or subtraction:

$$S_{Total} = \sqrt{S_1^2 + S_2^2 + \cdots S_k^2}$$

Propagation involving multiplication of division:

$$S_{Total} = \frac{RSD_{Total}}{100}$$

where:

$$RSD_{Total} = \sqrt{RSD_1^2 + RSD_2^2 + \cdots RSD_k^2}$$

samples) to calculate a measure of dispersion for the results (sample standard deviations or the relative standard deviations). The question is how to handle the variability of these independent variables. In other situations there may be a serial progression, and the first step involves a certain amount of error. The error would be compounded with the error associated with the second step in the procedure. This is further compounded with the third step and so forth until the last step in a procedure.

There are two methods for dealing with the propagation of error, and the choice depends on the mathematical process that takes place. For addition or subtraction (i.e., the previous serial example), the error term is based on the uncertainty measured by the variances of the independent measurements. For multiplication or division (e.g., surface area example), the error term is based on the relative uncertainty (RSD) of the independent measurements. The equations for these propagations are presented in Fig. 4.14.

For example, assume the standard deviation for a specific balance is 1.5 and the results of weighting 30 capsules is 201.07 ± 3.21 mg. The confounding of the two uncertainties is additive so the propagated error (calculated as the square root of the sum of the variances) would be a standard deviation of 3.54 mg. Additional information of the propagation and under different conditions is discussed by Taylor (1997).

References

Bolton S, Bon C (2004) Pharmaceutical statistics: practical and clinical applications, 4th edn. Marcel Dekker, Inc., New York, pp 159–161

De Muth JE (2014) Basic statistics and pharmaceutical statistical applications, 3rd edn. CRC Press, Boca Raton, FL, pp 158–159

FDA. CDER (1996) Q2B validation of analytical procedures: methodology, Rockville, MD

International Conference on Harmonization (2005) Q2(R1) Validation of analytical procedures: text and methodology

Kirk RE (1968) Experimental design: procedures for the behavioral science. Brooks/Cole Publishing Co., Belmont, CA, pp 9–11, 540–546

Taylor JR (1997) An introduction to error analysis: the study of uncertainties in physical measurements. University Science Books, Sausalito, CA, pp 45–92

USP 41-NF 36 (2018a) General Chapter <905> uniformity of dosage units. US Pharmacopeial Convention, Rockville, MD

USP 41-NF 36 (2018b) General Chapter <1010> analytical data—interpretation and treatment. US Pharmacopeial Convention, Rockville, MD

Young MJ et al (1983) Sample size nomograms for interpreting negative clinical studies. Ann Intern Med 99:248–251

Zar JH (2010) Biostatistical analysis, 5th edn. Prentice Hall, Upper Saddle River, NJ

Chapter 5
Multivariate Analysis: Tests to Identify Differences

Abstract This chapter explores tests where differences among levels of the independent variable are evaluated. Almost all inferential statistical tests (1) create a confidence interval or (2) calculate value from a ratio. In the former case, determination is made based on the values that appear within an interval. In the latter case, the ratio is based on the measure of what is being tested in the numerator and a measure of sample variability in the denominator. If the resultant statistic exceeds a critical value from a table, the hypothesis under test is rejected. Tests covered in this chapter include the t-tests (one-sample, two-sample, and paired tests), the analysis of variance (one-way, two-way, and N-way), the randomized complete block design test, and tests evaluating dichotomous proportions. If there is concern meeting the criteria for these tests, nonparametric alternative tests are available. Minitab applications to test for these differences are presented.

Keywords Analysis of variance · Multiple comparisons · One-tailed test · Paired data · Post hoc procedures · Randomized complete block design · t-test · Two-tailed test · Z-test of proportions

This chapter will focus on inferential statistics used to identify differences where (in all but a few univariate cases) there is at least one independent variable and one dependent variable. These would be among the most commonly seen tests for pharmaceutical analysis where the researcher controls discrete levels of the independent variable and the outcome is a continuous variable. Data can present as unpaired or paired data. The parametric tests for unpaired data include the t-test, Mann-Whitney test, analysis of variance, and Kruskal-Wallis test. If the sample data can be paired, appropriate tests include the paired t-test, Wilcoxon test, complete randomized block design, and Friedman test. In addition, tests for dealing with proportions will be discussed.

© American Association of Pharmaceutical Scientists 2019
J. E. De Muth, *Practical Statistics for Pharmaceutical Analysis*, AAPS
Advances in the Pharmaceutical Sciences Series 40,
https://doi.org/10.1007/978-3-030-33989-0_5

5.1   Types of Inferential Tests

As notes in Sect. 3.1, parametric statistics have two additional requirements in addition to the basic criteria for all tests, namely, good sampling and care to make sure each data point is independent of any other data point. The first of these requirements is that the population from which the sample is taken is assumed to be normally distributed (bell-shaped curve). Second is that the variances for each level of the independent variable are approximately equal (homogeneity of variance). The best estimate for these two requirements is to look at the sample data. Does the distribution in a histogram or dot plot for the sample appear to be symmetrical? Are the variances relatively close in size? A quick rule of thumb is the largest variance should not be more than twice the smallest variance. As noted in Chap. 3, statistical tests are available to test for normality and homogeneity. Many of the parametric tests have been employed for over 100 years. They are robust statistics and can handle slight to moderate deviations from the requirements of normality and homogeneity. Three factors can influence the results of these statistical tests, normality, homogeneity, and sample size. Each level of the independent variable should have approximately an equal number of data points. All the statistics in the book will allow for slight deviations in equal sample sizes, due to lost data points. However these differences should be minimal ($n_1 = 30$, $n_2 = 29$). Avoid comparisons where there are 10 observations in one group and 50 in the second. It is the one factor that the researcher can control.

5.1.1   One-Tailed Versus Two-Tailed Tests

There are two ways in which the type I error (α) can be distributed. In a *two-tailed test*, the rejection region is equally divided between the two ends of the sampling distribution. In this case both extremes of the distribution would be $\alpha/2$. For example, assume that a researcher wants to compare a new analytical method to a tradition method for analyzing the same product. Here there is an experimental method compared to a control (traditional) method. A two-tailed test would not predict that one is superior or better than the other, simply that a difference between the two methods exists. In this case the alternative hypothesis indicated that there is a different.

$$H_0 : \quad \mu_{experimental} = \mu_{control}$$
$$H_1 : \quad \mu_{experimental} \neq \mu_{control}$$

In contrast, a *one-tailed test* is a test of hypothesis in which the rejection region is placed entirely at one end of the sampling distribution. In the current example, assume researcher wants to prove that the new experimental method is superior to traditional control method:

$$H_0 : \mu_{experimental} \leq \mu_{control}$$
$$H_1 : \mu_{experimental} > \mu_{control}$$

If a one-tailed test is used, all the type I error is loaded on one side of the equation, and the decision rule with $\alpha = 0.05$ would be to reject the null hypothesis if $Z > Z_{CV}$ $(1 - \alpha)$. This is seen in Fig. 5.1 for 95% confidence, where the critical value for a two-tailed test is 1.96 (point at which the null hypothesis is rejected). However, for a one-tailed test with all the error loaded on one side, the critical value drops to 1.64. Using a one-tailed test, the researcher can prove the new method is superior to the existing method if the hypothesis is rejected. Minitab defaults to a two-sided test but in many cases allows the user to choose a one-tailed test under "Options" on the dialog boxes (Fig. 5.2). Selection choices for Minitab include "difference $\neq$ hypothesized difference" for a two-tailed test and "difference > hypothesized difference" or "difference < hypothesized difference" for one-sided tests.

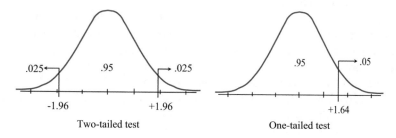

Fig. 5.1 Comparison of type I error distributions in a one-tailed and two-tailed test

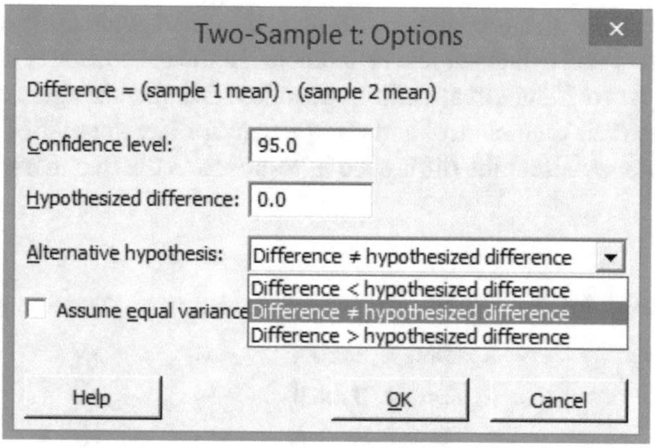

Fig. 5.2 Example of Minitab "Options" available for type I error

5.1.2   Tests for Paired Versus Unpaired Data

With an unpaired set of data, it is assumed that every measurement is completely independent of any other measurement (one of the critical assumptions noted in Chap. 1). The matched pair (or paired test) is used when complete independence does not exist between multiple samples, multiple time periods, or repeated measures. For example, assume a new piece of equipment is going to be tested to replace existing equipment. Several different products, from different batches, will be tested on each piece of equipment. The results of the study will be presented similar to Fig. 5.3. Different batches of drugs might be tested using each of the two pieces of equipment. Each row represents a level of the various drugs and batches being tested. The second and third columns are the actual outcomes. The last column is the difference between the first two columns per the independent variable. Traditionally the first measure (control) is subtracted from the second measurement (test). Therefore a positive difference represents a larger outcome on the second measure. In the case of a before and after (an intervention) design, the before measurement would be subtracted from the after results.

The potential problem is lack of independence between the "test" and "control" levels because the same batch is measured by both types of equipment. This is overcome because the research is not concerned with the individual scores but the difference score in the last column. With good science these difference scores will be independent of each other. The next question would be the sample size (n) and degrees of freedom. In this design since the researcher will only be concerned with the differences, the degrees of freedom would be $n - 1$ number of pairs or $n - 1$ number of samples tested under both conditions.

Two other examples of paired would include pretest and posttest design or crossover clinical trials. In the former case, the same individuals would take a pretest and posttest after some training intervention. It is assumed that the results on the posttest will be affected (not independently) by the pretest. The individual actually serves as a control. Therefore the test statistic is not concerned with differences between groups but actual individual subject differences. In the second case, volunteers may be exposed to two different treatments in a random order. Once again each volunteer would serve as their own control and compensate for any anomalies or concurring conditions that may affect the difference in response to the two therapies.

Fig. 5.3 Data design for paired data	Control	Test	d (Test – Control)
Drug A, Batch A	x_1	x'_1	$d_1 = (x'_1 - x_1)$
Drug A, Batch B	x_2	x'_2	$d_2 = (x'_2 - x_2)$
Drug B, Batch A	x_3	x'_3	$d_3 = (x'_3 - x_3)$
	...	...	
Drug K, Batch K	x_4	x'_n	$\underline{d_n = (x'_n - x_n)}$
			Σd

This paired test is more powerful than an unpaired test because each pair is serving as its own control and one would expect to get the same response for each equipment. With an unpaired test, both the old and new equipment would receive multiple random samples from the same batch, and each measurement would be assumed to be independent of any other measure. Examples of paired data discussed below will include the paired t-test, Wilcoxon test, the complete randomized block design, and the Friedman test.

5.1.3 Alternative Nonparametric Tests

Nonparametric tests were introduced in Sect. 3.8. Nonparametric statistical tests can be useful when dealing with small sample sizes or when the requirement of a normally distributed population cannot be met or assumed. Also called *distribution-free statistics*, these nonparametric statistical tests do not make any assumptions about the population distribution. One does not need to meet the requirements of normality or homogeneity of variance associated with the parametric procedures. Mostly developed in the late 1940s and 1950s, these nonparametric tests have been slow to gain favor in the pharmaceutical community but are currently being seen with greater frequency in the literature, often in parallel with their parametric counterparts.

The nonparametric tests are relatively simple to calculate. Their speed and convenience offer a distinct advantage over the parametric alternatives discussed in the this and future chapters. Therefore, as investigators we can use these procedures as a quick method for evaluating data. As seen is Sect. 3.8.1, nonparametric tests usually involve ranking or categorizing the data and by doing so decrease the accuracy of the information (changing from the raw data to a relative ranking). Doing so can obscure the true differences and make it difficult to identify differences that are significant. In other words, nonparametric tests require differences to be larger if they are to be found significant. We increase the risk of accepting a false null hypothesis (type II error). It may be to the researcher's advantage to tolerate minor doubts about normality and homogeneity associated with a given parametric test, rather than to risk the greater error possible with a nonparametric procedure.

When dealing with ordinal-dependent variable results, the nonparametric tests become the tests of choice (Daniel, p.16). As discussed in Chap. 2, units on an ordinal scale may not be equidistant and violate assumptions required for parametric procedures. Therefore the conversion from the initial ordinal scale to the relative positioning of a rank order scale would be the more appropriate statistical test.

In many nonparametric statistics, the median is used instead of the mean as a measure of central tendency. For a nonparametric test, the confidence interval is created around the median or comparing sample data to a hypothesized population median. To analyze differences between two or more discrete levels of the independent variable, the medians for each level are compared.

Nonparametric tests are particularly useful when there is a potential outlier (to be discussed in Chap. 8). Because it involves ranking data from 1 to N, the smallest and largest data point will still receive the ranks of 1 or N, respectively, regardless of how extreme those values might be. For example, assume the following numbers: 100, 100.3, 100.3, 100.4, 100.6, and 110.2. In this case 110.2 would seem different from the other five observations. However, when ranking the data, the value 110.2 would be converted to rank 6, and its difference from the other observations would be minimized. What if the largest value was 120.2 or even 150.2? The same rank of 6 would still be assigned. Thus, nonparametric statistics are not affected by a single outlier.

5.1.4 Minitab Applications (General Consideration)

Most parametric tests are located under Stats → Basic Statistics or Stats → ANOVA, while the nonparametric alternatives are at Stats → Nonparametric tests. Minitab defaults to a type I error of 0.05 and two-sided tests. These can be changed using "Options" on the dialog box (Fig. 5.2). Rarely is the hypothesized difference something other than the default of zero. As will be discussed in the two-sample t-test, some test will allow for a correction factor for slightly unequal variances, noted by the box for "Assume equal variances" or "Equal variance."

Referring back to Sect. 1.7.2 on how data can be presented in the worksheet, most tests require that each column is a variable ("Samples in one column"). For example, data presented in columns 1 through 3 in Fig. 5.4 show that each variable is in a separate column (site and percent) and characteristics to each sample are presented as a row. However some tests allow the option "Each sample is in its own column" that requires each level of independent variable be in a separate column. This layout is presented in columns 5 and 6 in Fig. 5.4 where each site of testing is in a separate column. Finally, some test will allow the entry of "Summarized data" (sample size, means, and standard deviations) from an existing report which can be

↓	C1-T	C2-T	C3	C4	C5	C6	C7
	Sample	Site	Percent		CRO	Pharma	
4	A4	CRO	101.3		98.6	103.7	
5	A5	CRO	100.6		101.7	99.7	
6	A6	Pharma	100.5		104.3	102.5	
7	A7	CRO	98.6		100.2	100.7	
8	A8	Pharma	103.7		99.4	104.3	
9	A9	Pharma	99.7		100.0	100.2	

Fig. 5.4 Various layouts for Minitab sample data

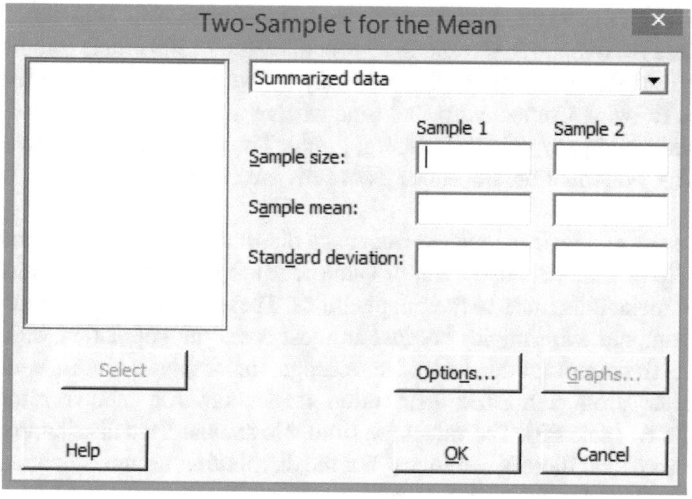

Fig. 5.5 Example of the main dialog box for summary data

entered for the inferential statistic (example of a two-sample t-test in Fig. 5.5). This last option is extremely helpful in reviewing papers or applying inferential statistics to information in summary reports.

5.2   One-Sample t-test

The one-sample t-test is a univariate statistic that is simply an extension of the one-sample Z-test for the common situations where the population standard deviation (σ) is unknown. The same type of confidence interval is created, but the reliability coefficient is based on a Student t-distribution, and the standard error term is expressed as a function of the sample standard deviation (S).

5.2.1   Dealing with Samples Where the Population Dispersion Is Unknown

The disadvantage with the formula for a one-sample Z-test is that it requires knowledge of the population standard deviation (σ) as discussed in Sect. 3.6. In most research, the population standard deviation is unknown, or at best a rough estimate can be made based on previous research (i.e., initial clinical trials or previous production runs). For larger sample sizes, the standard error of the mean will become more constant; therefore, the Z-test is accurate only for large samples. It would seem logical that for smaller sample sizes, a more conservative statistic would be required when the population standard deviation is unknown. This was noted and rationalized by William S. Gossett in a groundbreaking 1908 article (Student 1908). He published this work under the pseudonym "Student" because he worked for Guinness Brewing Company. At that time writing and publishing scientific papers were against company policy (Salsburg, p. 26). The distribution became known as the Student t-distribution and subsequent tests are called *Student t-tests* or simply *t-tests*.

The t-tests and their associated frequency distributions are used (1) to compare one sample to a known population or value or (2) to compare two samples to each other and make inferences to their populations. These are the most commonly used tests to compare two samples because in most cases the population variances are unknown. To correct for this lack of knowledge, the t-tables are used, which adjust the Z-values from a normal distribution to account for smaller sample sizes (Appendix B, Table B2). The values are similar to standardized distribution with the center at zero, and there is symmetry, but the distribution becomes flatter and more widely distributed as the sample size becomes smaller.

Please note several points about Table B2 that will be seen in other future tables discussed in this book. First, the left column is labeled as "df" for degrees of freedom, not sample size. In the case of a one-sample t-test, the degrees of freedom are $n - 1$. This is the same number of degrees of freedom that were used for calculating the sample standard deviation (Sect. 2.2.2). Second, the notation at the top of the table indicated that this is a table for a two-sided test; the values for each column represent $1 - \alpha/2$. Assume the researcher wants to be 95% confident that a sample is what is expected ($\bar{X} \sim \mu$) for a given product; then the acceptable type I error would 5% or $\alpha = 0.05$. Further assume it is a two-sided test (trying to identify a difference, not a prediction the sample is greater than or less than the expected results); therefore $\alpha/2 = 0.025$. Using this table the critical value for $1 - \alpha/2$ (in this case $1-0.025 = 0.975$) would be column labeled 0.975. Then the researcher would simply pick the reliability coefficient ($t_{df(1 - \alpha/2)}$) from the table and do the calculation in Fig. 5.6 using the sample mean and sample standard deviation. Lastly, note what the t-values are on Table B2 at infinity; they are the same as their corresponding Z-values in Table B1. Going up the table, as the sample size becomes smaller (fewer degrees of freedom), the reliability coefficient gets bigger, and the confidence interval will get larger and more conservative in its estimation of the population parameter μ, because less information is known about the population.

One use of one-sample t-test is to take sample means and make a statement about a population. Based on a sample mean ($\bar{X}$) and standard deviation (S), with a sample size of n, the true population falls somewhere within an interval:

$$\text{Lower Limit} < \mu < \text{Upper Limit}$$

The test does not specify an exact location for the population mean. It simply indicates that the population mean is located somewhere between the two limits with a certain amount of confidence.

One sample t-test
  (confidence interval):

$$\mu = \bar{X} \pm t_{(1-\alpha/2)} \times \frac{S}{\sqrt{n}}$$

One sample t-test (ratio method):

$$t = \frac{\bar{X} - \mu}{\frac{S}{\sqrt{n}}}$$

Two sample t-test (confidence interval):

$$\mu_1 - \mu_2 = (\bar{X}_1 - \bar{X}_2) \pm t_{n_1+n_2-2}(1 - \alpha/2) \times \sqrt{\frac{S_p^2}{n_1} + \frac{S_p^2}{n_2}}$$

where (pooled variance):

$$S_p^2 = \frac{(n_1 - 1)S_1^2 + (n_2 - 1)S_2^2}{n_1 + n_2 - 2}$$

Two sample t-test (ratio method):

$$t = \frac{\bar{X}_1 - \bar{X}_2}{\sqrt{\frac{S_p^2}{n_1} + \frac{S_p^2}{n_2}}}$$

Paired t-test (confidence interval):

$$\mu_d = \bar{X}_d \pm t_{n-1}(1 - \alpha/2) \times \frac{S_d}{\sqrt{n}}$$

Paired t-test (ratio method):

$$t = \frac{\bar{X}_d}{\frac{S_d}{\sqrt{n}}}$$

where: d is the difference
  between pairs

$$\bar{X}_d = \frac{\sum d}{n} \qquad S_d^2 = \frac{n(\sum d^2) - (\sum d)^2}{n(n - 1)}$$

Fig. 5.6 Equations for Student t-tests

A second approach to using the one-sample t-test could be used to determine if a sample mean is equal to a suspected population mean:

$$H_0 : \quad \bar{X} = \mu$$

$$H_1 : \quad \bar{X} \neq \mu$$

If resulting confidence interval includes the value of μ, then the null hypothesis cannot be rejected, and the best guess is that the sample has a similar mean. If the confidence interval includes only possibilities that are greater than or less than the hypothesized population mean, then the null hypothesis is rejected in favor of the alternative hypothesis.

A second approach would be to use a parallel ratio method for calculating a *t*-statistic (Fig. 5.6). As with all statistical ratios, the factor being tested is in the numerator (in this case the sample mean), and some measure of error is in the denominator. The resultant statistic would be compared to the critical values in Table B2 in Appendix B. Fortunately Minitab simplifies the process by reporting the *t*-value and subsequent *p*-value. If the *p*-value is less than 0.05, reject the null hypothesis with at least 95% confidence in that decision. Use of these tests will be presented in an example in Sect. 5.2.5.

5.2.2 Sample Size and Statistical Power

As noted in Sect. 4.3, often the researcher will want to know if the sample size is adequate to determine if the sample result are significant differences from an expected population mean and/or if there is significant power to reject a false null hypothesis. Minitab can perform these calculations, and the statistics for these determinations are described by Zar (pp. 107–108).

5.2.3 Nonparametric Alternative

If the population is known or suspected to have a distribution other than a symmetrical normal distribution, then nonparametric alternative approaches are available. One approach is the sign test which is available on Minitab. Based on sample data, the null hypothesis is that the population median from which the sample was selected is equal to some hypothesized median. The alternative hypothesis is that they are not equal:

$$H_0 : \quad \text{median}\left(M_i\right) = \text{hypothesized median}\left(M_0\right)$$
$$H_1 : \quad \text{median}\left(M_i\right) \neq \text{hypothesized median}\left(M_0\right)$$

The one-sample *sign test* can be used to estimate the population median based on sample data, create a confidence interval around the median, or compare sample results with a predetermined hypothesized population median. It is called the sign test because data are converted to plus (+) and minus (−) signs depending on where they are larger or smaller than the hypothesized median. It is one of the oldest nonparametric procedures, reported as early as 1710 by British physician John Arbuthnott (Hollander and Wolfe, p. 28).

   The sign test is an example of an *exact method*, meaning the actual *p*-value is the computed statistic, and it does not involve an intermediate statistic (e.g., *t*-value). With Minitab, for up to a sample size of 50, the software uses the binomial equation. Scores equal to the hypothesized median are removed, and probability of being greater than the hypothesized median is 0.50. For a sample size greater than 50, Minitab uses a normal approximation of the binomial equation. These computations are described by Conover (pp. 157–164). The simplest interpretation is to evaluate the *p*-value. If the *p*-value is less than 0.05, then reject the null hypothesis. If greater than or equal to 0.05, one is unable to reject the null hypothesis.

5.2.4   Minitab Applications

With Minitab it is possible to do the one-sample t-test and calculate the corresponding sample size and power determinations. In addition the sign test is available for non-normal distributions. Steps for performing these tests are presented below.

5.2.4.1   One-Sample t-test

Procedure	Stats ➔ Basic Statistics ➔ One-Sample t-test
Data input	Select the source of information at the top of the dialog box: (1) data where each column is a variable on the worksheet; or (2) summary data is available for the sample size, mean and standard deviation.
Hypothesized mean	Without a hypothesized mean, Minitab will calculate a confidence interval. With a hypothesized mean, it will calculate both a confidence interval and a *t*-statistic with corresponding *p*-value.
Options	Automatic default for a 95% confidence interval for a two-tailed test. These can be changed to a different degree of type I error if required or a one-tailed test is desired.
Graphs	None by default can request histogram, individual value plot (dot plot), or box plot. Each will display the confidence interval below the graphic.
Report	Without a hypothesized mean, a confidence interval is produced. With a hypothesized mean provided, it will create both a confidence interval and *t*-statistic with corresponding *p*-value.
Interpretation	Without a hypothesized mean, the confidence interval produced the limits within which the population mean should fall with a given degree of confidence. With a hypothesized mean, reject the null hypothesis if hypothesized mean falls outside the confidence interval or if the *p*-value is less than 0.05.

5.2.4.2 Sample Size and Power Determination

Procedure	Stat → Power and Sample Size → One-Sample t-test
Data input	Standard deviation is required and two of the following three: (1) sample size, (2) detectable difference, and (3) power. Program will solve for the third item not entered. Multiple values can be selected for any one of the two input items (e.g., sample sizes of 10, 20, 30, and 40).
Options	Automatic default for a 95% confidence interval for a two-tailed test. These can be changed to a different degree of type I error if required or a one-tailed test is desired.
Graphs	By default will produce a power curve, which can be turned off under "Graph." Multiple sample sizes can be added to create a graph with multiple curves and power points.
Report	Results for the third item not entered (e.g., if sample size and detectable difference are entered, results will report the power). Multiple entries (e.g., sample size) will report results for each sample size entered.
Interpretation	Helps identify the power for different sample sizes or different detectable differences or the required sample size for a desired power and detectable difference.

5.2.4.3 Nonparametric Alternative: Sign Test

Procedure	Stats → Nonparametric → One-Sample Sign
Data input	Selected column from the worksheet.
Options	Can be run with or without a hypothesized median. Without a hypothesized median, it will automatic default for a 95% confidence interval for a two-tailed test, but the confidence interval can be changed. If a hypothesized median is provided, the confidence interval can be changed, and one-tailed alternatives can be selected.
Results	Without a hypothesized median, only a confidence interval will be reported. With a hypothesized median, both a confidence interval and p-value are reported.
Interpretation	Without a hypothesized median, the resulting confidence interval will represent the limits within which the population median will fall with a given degree of confidence. With a hypothesized median, the easiest interpretation is to look at the reported p-value. If the p is less than 0.05, reject the null hypothesis in favor of the alternative hypothesis; if not, fail to reject the null hypothesis.

One-sample t-test:

N	Mean	StDev	SE Mean	95% CI for μ
10	102.45	3.86	1.22	(99.69, 105.21)

μ: mean of %LC

Sign test:

Sample	N	Median
%LC	10	102.05

Sample	CI for η	Achieved Confidence Position
%LC	(99.1, 105.1)	89.06% (3, 8)
	(98.8604, 105.477)	95.00% Interpolation
	(98.4, 106.2)	97.85% (2, 9)

Fig. 5.7 Minitab output for confidence intervals for analysis with one-sample t-test and nonparametric alternatives

5.2.5   Example

A pharmaceutical manufacturer is developing a chewable gel product and randomly samples ten gels from a scale-up run. The content results are 99.1, 105.1, 103.5, 110.0, 100.2, 106.2, 98.4, 101.3, 102.8, and 97.9 percent label claim, with a mean of 102.45% and standard deviation of 3.86%. The one-sample t-test creates a confidence interval of 99.69 to 105.21 (Fig. 5.7). Therefore, it can be stated with 95% confidence that the population mean for the scale-up run is somewhere between 99.7% and 105.2% label claim. Alternatively, the results can be compared to a target goal of 100% label claim. Using these sample results, the null hypothesis that the estimated population equals 100% cannot be rejected because the *t*-statistic is 2.01 with a corresponding *p*-value of 0.076 (Fig. 5.8). In other words, if the scientist wanted to reject the null hypothesis, she would need to live with a 7.6% chance of being wrong. Rejecting the null hypothesis is unacceptable, if the required confidence level is 95%.

With such a small sample size, the researcher is concerned that there might not be adequate statistical power. Her goal is to have at least 80% power ($p_{1-\beta} = 0.80$) to detect at least a 5% difference from the target of 100%. Using Minitab she is able to determine that she had more than adequate power (0.952 in Fig. 5.9) based on ten observations to identify a difference from the population mean of 5%. In this case

One-sample t-test:

Null hypothesis	H_0: μ = 100
Alternative hypothesis	H_1: $\mu \neq$ 100

t-value	P-Value
2.01	0.076

Sign test:

Null hypothesis	H_0: η = 100
Alternative hypothesis	H_1: $\eta \neq$ 100

Sample	Number < 100	Number = 100	Number > 100	P-Value
%LC	3	0	7	0.344

Fig. 5.8 Minitab output for comparison to an estimated population median (100% label claim) with a one-sample t-test and nonparametric alternatives

Power given 10 observations to detect a 5% difference:

1-Sample t Test
Testing mean = null (versus ≠ null)
Calculating power for mean = null + difference
α = 0.05 Assumed standard deviation = 3.86

Difference	Sample Size	Power
5	10	0.952619

Required observations for power of 0.80 to detect a 5% difference:

Difference	Sample Size	Target Power	Actual Power
5	7	0.8	0.813364

Fig. 5.9 Minitab output for sample size and power determinations from the one-sample t-test example

the sample standard deviation (S) was used as the best estimate of σ. For this example to have a desired power of 0.80 to detect a 5% difference, she would need a sample size of only seven observations (lower portion of Fig. 5.9).

Because of the small sample size and early in development, the researcher is concerned that the population of chewable gels may not be normally distributed and decides to also run a nonparametric procedure. As seen in Figs. 5.7 and 5.8, the results for the sign test are similar to those for the one-sample t-test. The p-value is greater, and the confidence interval is slightly wider, but the researcher would make the same decision. With either test she would fail to reject the null hypothesis that $\mu = 100\%$. It is important to note that she does not *prove* the null hypothesis, but simply does not have enough evidence to reject that hypothesis.

5.3 Two-Sample t-test

The most common use for the t-test is for comparing two levels of a discrete independent variable with a continuous dependent variable. For example, a control and experiment method are compared with the results expressed as weight or percent.

Similar to the one-sample case, there are two approaches that can be taken in performing a two-sample t-test: (1) establish a confidence interval for the population differences or (2) use a ratio method and compare test results to a critical value. The same example for the two-sample t-test will be used to illustrate these two methods of hypothesis testing. There are two possible hypotheses that are identical statements.

	Confidence interval	Critical value
The population means are the same :	$H_0 : \mu_1 - \mu_2 = 0$	$H_0 : \mu_1 = \mu_2$
The population means are different :	$H_1 : \mu_1 - \mu_2 \neq 0$	$H_1 : \mu_1 \neq \mu_2$

Note that the hypotheses are saying the same thing. For the null hypothesis, μ_1 and μ_2 are the same (there is no difference), and the alternative (mutually exclusive and exhaustive) statement is that they are different. Either the confidence interval or ratio method will produce the same results. Minitab will report the results from both approaches.

In this test the number of degrees of freedom is defined as $n_1 + n_2 - 2$ or $N - 2$. If one needs the reliability coefficient for a confidence interval

$$\text{Population} = \begin{pmatrix} \text{Estimate} \\ (\text{sample data}) \end{pmatrix} \pm \begin{pmatrix} \text{Reliability} \\ \text{Coefficient} \end{pmatrix} \times \begin{pmatrix} \text{Standard} \\ \text{Error} \end{pmatrix}$$

or the critical value to determine of the t-statistic was significant

$$\text{Test Statistic} = \frac{\text{Sample Estimate}}{\text{Standard Error}}$$

the same number would be used. This would be selected from Table B2 (Appendix B) for $N - 2$ degrees of freedom and a level of confidence selected by the researcher, usually 95%. The column in the table would be 0.975 ($1 - \alpha/2$). One could write a decision rule for determining significance:

With $\alpha = ((\text{researcher's choice}))$
reject H_0 if t-value is greater than ((absolute critical value Table B2)).

For the two-sample t-test, the standard error term was a pooled variance (S_p^2 in Fig. 5.6). This is a weighted average of the two variances. The best estimate of the population $\mu_1 - \mu_2$ will be the sample results of $\overline{X}_1 - \overline{X}_2$.

Determination of significance is based on whether or not zero falls within the confidence interval. If zero is within the interval, the null hypothesis $\mu_1 - \mu_2 = 0$ cannot be rejected. If all the results fall to the positive or negative side, then zero is not a possible result, and the null hypothesis can be rejected in favor of the alternative hypothesis $\mu_1 - \mu_2 \neq 0$. The second approach for doing a two-sample t-test is to create the ratio seen in Fig. 5.6. This ratio approach is used in many statistical tests with the general formula listed above. The result is compared to the critical value (the value on the t-table). If the absolute value of the test statistic is greater than the critical value, the null hypothesis is rejected. If the absolute value is less than the critical value, then the sample difference is due to chance alone, and the null hypothesis cannot be rejected.

As will be seen with this test and most of the remaining inferential statistics, computer program will present not only the test statistic result but also a p-value (type I error). So reference to a critical value and tables become less important. The determination is to reject the null hypothesis if the p-value is less than 0.05, reflecting at least 95% confidence in rejecting the hypothesis. This will be illustrated in the example below.

5.3.1   Correction for Unequal Sample Sizes and Unequal Variances

In the ideal world, both levels of the independent discrete variable would have an equal number of observations. Unfortunately this often does not happen (samples may be lost or contaminated). Because of the homogeneity of variance requirement, it would be ideal if both levels of the independent variable have the same identical sample variance. In most cases this will not happen, even though the required criteria are to have relatively similar dispersions (homogeneity of variance). Increases in the difference in sample sizes and/or increased differences between the variances with stress the two-sample t-test. A correction factor, *Welch-Satterthwaite solution* (or simply the *Satterthwaite solution*), can be used to downgrade the number of degrees of freedom for increasing differences (Satterthwaite 1946). The downregulation or decrease in the degrees of freedom will increase the size of the confidence interval and increase the p-value (Fig. 5.10). This corrective solution is automatically applied in Minitab unless the researcher chooses "Assume equal variances" under "Options" in the dialog box.

Fig. 5.10 Satterthwaite
correction for unequal
sample sizes and unequal
variances

$$df = \frac{\left(\frac{S_1^2}{n_1} + \frac{S_2^2}{n_1}\right)^2}{\frac{\left(\frac{S_1^2}{n_1}\right)^2}{n_1} + \frac{\left(\frac{S_2^2}{n_2}\right)^2}{n_2}}$$

5.3.2 Sample Size and Statistical Power

Similar to the one-sample t-test, the scientist may want to know if the sample size is adequate to determine if significant differences exist between these two levels of the independent variable and/or if there is adequate power to reject the null hypothesis. Again, Minitab can perform these calculations, and the statistics for these determinations are provided by Zar (pp. 131–136).

5.3.3 Nonparametric Alternative

A nonparametric alternative to the two-sample t-test in the *Mann-Whitney test*. As with other nonparametric procedures, it involves ranking the data across the two levels of the independent variable (Mann and Whitney 1947). The Mann-Whitney test has numerous synonyms, including the Mann-Whitney U, two-sample Wilcoxon rank sum test, and the Wilcoxon-Mann-Whitney (WMW) test. In the parametric t-test, the normality was assumed, and the null hypothesis was $\mu_1 = \mu_2$. But in nonparametric procedures, the distribution of population data is not a concern, and the median is used as the measure of center for these procedures. The hypotheses are much simpler:

H_0 :   Samples are from the same population

H_1 :   Samples are drawn from different populations

There are two assumptions for performing the Mann-Whitney test: (1) the populations from which the samples are taken have the same shape; and (2) observations are independent of each other. If the populations are normally distributed, this test is slightly less powerful than the two-sample t-test, and the confidence interval will be wider. The equation for calculating this test is discussed by Daniels (pp. 82–87).

5.3.4 Minitab Applications

This test is for comparing two levels of the discrete independent variable and a continuous dependent variable. Minitab can perform the two-sample t-test, corresponding sample size, and power determinations, as well as the Mann-Whitney U test. Steps for performing these tests are presented below.

5.3.4.1 Two-Sample t-test

Procedure   Stats → Basic Statistics → 2-Sample t

Data input   Select the source of information at the top of the dialog box: (1) data where each column is a variable on the worksheet ("Samples" is the dependent variable and "Sample IDs" is the independent variable); (2) each level of the independent variable is in a different column in the worksheet; or (3) summary data is available for the sample size, mean, and standard deviation.

Options   Automatic default for a 95% confidence interval for a two-tailed test. These can be changed to a different degree of type I error if required or a one-tailed test is desired. Also, by default it is assumed that the variances are not equal and the Satterthwaite solution will be used. Clicking on "Assume equal variances" will remove this correction factor.

Graphs   None by default can request an individual value plot (dot plot) or box plot.

Report   Both a confidence interval and a t-statistic with corresponding p-value for the difference between the two means are presented. Also reports the sample means and standard deviations.

Interpretation   With a hypothesized mean of zero, reject the null hypothesis if zero falls outside the confidence interval or if the p-value is less than 0.05.

5.3.4.2 Sample Size and Power Determination

Procedure   Stat → Power and Sample Size → 2-Sample t

Data input   Standard deviation is required and two of the following three: (1) sample size, (2) detectable difference, and (3) power. Program will solve for the third item not entered. Multiple values can be selected for any one of the two input items (e.g., sample sizes of 10, 20, 30, and 40).

Options   Automatic default for a 95% confidence interval for a two-tailed test. These can be changed to a different degree of type I error if required or a one-tailed test is desired.

Graphs   By default will produce a power curve, which can be turned off under "Graph." Multiple sample sizes can be added to create a graph with multiple curves and power points.

Report   Results for the third item not entered (e.g., if sample size and detectable difference are entered, results will report the power). Multiple entries (e.g., sample size) will report results for each sample size entered.

Interpretation   Helps identify the power for different sample sizes or different detectable differences or the required sample size for a desired power and detectable difference.

5.3.4.3 Nonparametric Alternative: Mann-Whiney Test

Procedure   Stats → Nonparametric → Mann-Whitney

Data input   Results for each level on the independent variable *must* be in separate columns.

Options   On the dialog box, defaults for 95% confidence and a two-tailed test can be changed if desired.

Results   Reports a confidence interval and a w-value with associated p-value. Also reports the median for each level of the independent variable.

Interpretation   The easiest interpretation is to look at the reported p-value. If the p is less than 0.05, reject the null hypothesis in favor of the alternative hypothesis; if not, fail to reject the null hypothesis. A confidence interval that does not include zero as a possible outcome also would result in the rejection of the null hypothesis of no difference.

5.3.5 *Examples*

A foreign manufacturer is developing a nicotine patch for commercial release. At one point the manufacturer is comparing two trial lots for adhesion. The results are listed in Table 5.1. Is there a significant difference between these two most recent trial lots base on their adhesive quality (g/part)?

Note that the sample sizes are not equal ($n_i = 20$ and $n_2 = 18$). No problem. All the equations presented in this book and the Minitab applications can handle unequal cell sizes (levels of the independent variable) unless otherwise noted. In this case the pooled variance in Fig. 5.6 weights the variances based on their respective sample size.

The results of the Minitab evaluation of the data are presented in Figs. 5.11 and 5.12. With the default of 95% confidence, the two-sample t-test results in a confidence interval from −469 to −51 g/part. Since zero is not within the interval, the null hypothesis that the two lots have equal adhesion can be rejected and determined that Lot B has significantly greater adhesive qualities than Lot A (based on the difference in sample means) somewhere between 51 and 469 g/part. Viewing the parallel approach of the ratio method for the two-sample t-test, the results are $t = -2.52$, $p = 0.016$. Once again, the null hypothesis can be rejected with the chance of being wrong of only 1.6% or a confidence level of 98.4%. However, note that if the

Table 5.1 Samples from two lots measuring patch adhesion (g/part)

Lot A		Lot B	
2107.1	2353.1	2357.2	2912.4
2050.4	2505.8	2392.7	2681.4
1947.6	2782.0	2082.8	2648.0
1986.0	2655.7	2224.9	2962.0
2081.3	2486.3	2452.7	2946.6
2108.8	2215.5	2495.5	2715.9
2286.2	2685.8	2681.6	2514.7
2047.5	2785.3	2274.4	2889.0
1819.0	2828.5	1939.4	2929.0
1878.9	2418.0		
Mean = 2301.44		Mean = 2573.12	
Standard deviation = 324.66		Standard deviation = 315.31	
$n = 20$		$n = 18$	

Fig. 5.11 Minitab output for confidence intervals for analysis with a two-sample t-test and nonparametric alternative

Two-sample t-test:

Sample	N	Mean	StDev	SE Mean
Lot A	20	2301	325	73
Lot B	18	2561	310	73

Difference	95% CI for Difference
-260	(-469, -51)

Mann-Whitney test:

Sample	N	Median
Lot A	20	2250.85
Lot B	18	2581.35

Difference	CI for Difference	Achieved Confidence
-268.55	(-505.1, -35.3)	95.15%

researcher wanted to be 99% confident in his decision, he would fail to reject the hypothesis because the type I error rate is greater the 1%.

Assume that the researcher wants to be able to detect a 5% difference between the two batches if that difference exists and would like to have at least 80% power to do so. In this case, based on all the data collected, 5% of a grand mean of 2424.45

Fig. 5.12 Minitab output
for the ratio method for
analysis with a two-sample
t-test and nonparametric
alternative

Two-sample t-test:

| Null hypothesis | H_0: $\mu_1 - \mu_2 = 0$ |
| Alternative hypothesis | H_1: $\mu_1 - \mu_2 \neq 0$ |

t-value	DF	P-Value
-2.52	35	0.016

Mann-Whitney test:

| Null hypothesis | H_0: $\eta_1 - \eta_2 = 0$ |
| Alternative hypothesis | H_1: $\eta_1 - \eta_2 \neq 0$ |

W-Value	P-Value
311.00	0.022

Fig. 5.13 Minitab output
for sample size and power
determinations from the
two-sample t-test example

Power given 18 observations to detect a 5% difference :

2-Sample t Test
Testing mean 1 = mean 2 (versus $\neq$)
Calculating power for mean 1 = mean 2 + difference
$\alpha = 0.05$ Assumed standard deviation = 339.97

| | Sample | |
Difference	Size	Power
121.22	18	0.180057

The sample size is for each group.

Required observations for power of 0.80 to detect a 5% difference:

| | Sample | Target | |
Difference	Size	Power	Actual Power
121.22	125	0.8	0.801772

The sample size is for each group.

is 121.22 g/part, and the pooled standard deviation (339.97 g/part) would be the best estimate of the population standard deviation. Therefore, using the smaller sample size ($n_2 = 18$), the power would only be 0.180 (Fig. 5.13). A sample size of 125 observations per group would be needed to provide a power of 0.80.

Similar results would be seen if the researcher decided to use the nonparametric alternative. The Mann-Whitney test results for a confidence interval that does not include zero (Fig. 5.11) and a test statistics with *p*-value equal to 0.022 (Fig. 5.12,

Table 5.2 Results of analysis of the same drug at two different facilities

	Manufacturer	Contract lab
	101.1	96.5
	100.6	101.1
	98.8	99.1
	99	98.7
	98.7	97.8
	100.8	99.5
Mean =	99.833	98.783
SD =	1.124	1.699
n =	6	6

Fig. 5.14 Minitab output for the comparison of results from two laboratories using the ratio method for the two-sample t-test and nonparametric alternative

Two-sample t-test:

Null hypothesis	$H_0: \mu_1 - \mu_2 = 0$
Alternative hypothesis	$H_1: \mu_1 - \mu_2 \neq 0$

t-value	DF	P-Value
1.34	9	0.212

Mann-Whitney test:

Null hypothesis	$H_0: \eta_1 - \eta_2 = 0$
Alternative hypothesis	$H_1: \eta_1 - \eta_2 \neq 0$

Method	W-Value	P-Value
Not adjusted for ties	45.00	0.378
Adjusted for ties	45.00	0.377

slightly higher than the t-test results), providing the same outcome and statistical determination as the two-sample t-test.

For a second example, samples are taken from a specific batch of drug and randomly divided into two groups of tablets. One group is assayed by the manufacturer's own quality control laboratories. The second group of tablets is sent to a contract laboratory for identical analysis. Is there a significant difference between the results generated by the two labs? In this case the independent variable is the laboratory doing the testing, and the dependent variable is the resulting percent label claim. The results of the study are presented in Table 5.2 and Minitab results in Fig. 5.14. The results is a failure to find a significant difference between the results from the two laboratories ($t = 1.34$, $p = 0.212$). Similar results were seen with the Mann-Whitney test ($W = 45.00$, $p = 0.377$).

5.4   Paired t-test

The t-distribution also can be used for paired data. In the case of the paired t-test, it measures differences created by the sample at two points (e.g., before and after, under two different test conditions, with two different technicians). Sometimes referred to as *repeated measures* procedures, the test is used when complete independence does not exist between the two methods, individuals or time periods. For example, in a pretest-posttest instructional research design, where the same individual takes both tests, it is assumed that the results on the posttest will be affected (not independent) by the pretest. The individual actually serves as a control. Therefore the test statistic is not concerned with differences between groups but actual individual subject differences. The hypotheses are associated with the mean difference in the population based on sample data:

$$H_0 : \mu_d = 0$$
$$H_1 : \mu_d \neq 0$$

Here the best estimate of the population mean difference is the mean difference between the sample pairs. As with the previous two tests, the results can be evaluated as either a confidence interval or ratio method (Fig. 5.6). In this case the number of degrees of freedom is $n - 1$ pairs of data. For the confidence interval, the null hypothesis cannot be rejected if zero falls with the interval. For the ratio method, a $p < 0.05$ would result in rejection of the null hypothesis.

5.4.1   Sample Size and Statistical Power

Once again sample size and statistical power may be important to the researcher in determining if significant difference exists between the paired measures. Minitab can perform these calculations, and the statistics for these determinations are presented by Zar (pp. 161).

5.4.2   Nonparametric Alternative

A nonparametric alternative to the paired t-test is the *Wilcoxon signed-rank test*. Frank Wilcoxon was a chemist with American Cyanamid and Lederle Laboratories and developed several nonparametric procedures during the 1940s (Salsburg, p. 161). Also referred to as the Wilcoxon matched pairs test, it evaluates the magnitude of the differences from that media. The formula ranks the differences where (1) zero differences are excluded; (2) differences are ranked together based on magnitude regardless of whether the difference is positive or negative; and (3) the statistic is based on the sum of ranks associated with the sign (positive or negative) that occurs with least frequency. The formula and calculations for Wilcoxon signed-rank test are presented in Hollander and Wolfe (pp. 126–131).

5.4.3   Minitab Applications

These tests are for two levels of an independent variable that is paired and a continuous dependent variable. Minitab has options for the paired t-test, corresponding sample size and power determinations, and the nonparametric Wilcoxon signed-rank test. Steps for performing these tests are presented below.

5.4.3.1   Paired t-test

Procedure	Stats → Basic Statistics → Paired t
Data input	Type selected on dialog box: (1) the two levels of the independent variable on different columns from the worksheet (paired across rows) or (2) summary data with the sample size, mean difference, and standard deviation difference.
Options	Automatic default for a 95% confidence interval for a two-tailed test. These can be changed to a different degree of type I error if required or a one-tailed test is desired.
Graphs	None by default can request a histogram, individual value plot, or box plot of the differences. Confidence intervals will be displayed with each of these graphic.
Report	Both a confidence interval and a t-statistic with corresponding p-value for the difference between the paired samples are reported. Also presented are the sample means and standard deviations for both levels of the independent variable as well as the mean difference and standard deviation of the difference.
Interpretation	Reject the null hypothesis of no difference if zero falls outside the confidence interval or if the p-value is less than 0.05.

5.4.3.2   Sample Size and Power Determination

Procedure	Stat → Power and Sample Size → Paired t
Data input	Standard deviation is required and two of the following: (1) sample size, (2) detectable difference, and (3) power. Program will solve for the third item not entered. Multiple values can be selected for any one of the two input items (e.g., sample sizes of 10, 20, 30, and 40).
Options	Automatic default for a 95% confidence interval for a two-tailed test. These can be changed to a different degree of type I error if required or a one-tailed test is desired.
Graphs	By default will produce a power curve, which can be turned off under "Graph." Multiple sample sizes can be added to create a graph with multiple curves and power points.

Report   Results for the third item not entered (e.g., if sample size and
                      detectable difference are entered, results will report the power).
                      Multiple entries (e.g., sample size) will report results for each
                      sample size entered.
Interpretation   Helps identify the power for different sample sizes or different
                      detectable differences or the required sample size for a desired
                      power and detectable difference.

5.4.3.3   Nonparametric Alternative: Wilcoxon Signed-Rank Test

Procedure   Stats ➔ Nonparametric ➔ 1-sample Wilcoxon
Data input   A column must be created for the differences between the indi-
                      vidual pairs of data. The column of differences is selected for the
                      statistical test.
Options   On the dialog box, the defaults of 95% confidence and a two-
                      tailed test can be changed if desired.
Results   Wilcoxon test produced a median and confidence interval.
Interpretation   If the confidence interval includes a zero difference, the null
                      hypothesis cannot be rejected.

5.4.4   Examples

Two different methods are used to evaluate the contents of a raw material used in a
manufacturing process. Samples from ten different batches are tested using both analytical methods (Table 5.3). Is there a significant difference in the results comparing

Table 5.3 Comparison of different batches using two analytical methods

Batch	Method A	Method B	d	d^2	Rank
1	17.2	17.8	0.6	0.36	7.5
2	17.6	17.2	−0.4	0.16	6[a]
3	19.3	19.1	−0.2	0.04	3[a]
4	17.6	17.9	0.3	0.09	5
5	18.3	18.9	0.6	0.36	7.5
6	17	18.1	1.1	1.21	9
7	17.7	17.6	−0.1	0.01	1[a]
8	18.1	18.1	0	0	…
9	18.3	18.1	−0.2	0.04	3[a]
10	17.6	17.8	0.2	0.04	3
		Sum =	1.9	2.31	

[a]For nonparametric tests these are the ranks associated with the least frequent sign (+)

Paired t-test (confidence interval):

Sample	N	Mean	StDev	SE Mean
Method B	10	18.060	0.568	0.180
Method A	10	17.870	0.660	0.209

			95% CI for
Mean	StDev	SE Mean	μ_difference
0.190	0.465	0.147	(-0.143, 0.523)

μ_difference: mean of (Method B - Method A)

Paired t-test (ratio method):

Null hypothesis	H_0:μ_difference = 0
Alternative hypothesis	H_1: μ_difference ≠ 0

t-value	P-Value
1.29	0.229

Wilcoxon test:

Sample	N	Median	CI for η	Achieved Confidence
Analysis	10	0.2	(-0.15, 0.55)	94.72%

Fig. 5.15 Minitab output for the paired t-test for two analytical methods and nonparametric alternative

these two methods? Assuming that the population results are normally distributed for the content of this raw material, a paired t-test can be used since multiple batches are tested using both methods. In this case, as seen in Fig. 5.15, the researcher discovers that there is no significant difference between the two methods ($t = 1.29, p = 0.229$). Alternatively a confidence interval could be calculated (-0.143 to $+0.523$) with 95% confidence. Since zero is a possible outcome, it can be determined that there is no significant difference between the two methods. Once again, the results do not prove that the two methods were identical, only that no statistically significant difference was detected.

After the study the researcher wanted to determine if there was adequate power to identify a 20% difference between the two methods. With the sample standard deviation of the difference 0.465 and a 20% difference based on the mean difference of 0.038 (0.190×0.20), the result was an estimated power of only 0.056 (Fig. 5.16). In order to achieve a desired power of 0.80, a total of 1178 batches or samples would have been required!

Fig. 5.16 Minitab output for sample size and power determinations from the paired t-test example with two analytical methods

Power given 10 pairs to detect a 20% difference :

	Sample	
Difference	**Size**	**Power**
0.038	10	0.0561941

Required observations for power of 0.80 to detect a 20% difference:

	Sample	Target	
Difference	**Size**	**Power**	**Actual Power**
0.038	1178	0.8	0.800262

Table 5.4 Results of samples taken from the same run for two sides of a tablet press

Times	Side 1	Side 2
0:30	107.5	105.1
1:00	107.7	101.6
1:30	103.7	106
2:00	100.7	103.5
2:30	105.7	105.9
3:00	106.1	104.3
3:30	106.3	103.7
4:00	104.6	105.7
4:30	104.5	105.3
5:00	104.4	104.4
5:30	106.5	105.1
6:00	102.7	101.6

Using the nonparametric Wilcoxon signed-rank test, the results also include a confidence interval with zero (−0.15 to +0.55) and a failure to reject the null hypothesis that there is no population difference because zero is a possible outcome (Fig. 5.15).

As a second example, samples are taken at various times from both sides of the tablet press during the production of core tablets that will be eventually enteric coated. The results are seen in Table 5.4. Is there a significant difference in the percent of active ingredient between the two sides of the tablet press? In this case the design represents different times measured twice, by each side of the tablet press. Independence does not exist between the paired measurements; therefore the paired t-test is the most appropriate inferential statistic. The Minitab results are presented in Fig. 5.17. Clearly there was no significant difference in the analyses by the two sides of the tablet press ($t = 0.97$, $p = 0.353$). To conclude that there was a difference, the researcher would need to be willing to accept a 35% chance of being wrong. Certainly much greater than an acceptable type I error of 5%. The Wilcoxon test created a confidence interval with zero clearly within the interval, also resulting

Paired t-test (ratio method):

Null hypothesis H_0: µ_difference = 0
Alternative hypothesis H_1: µ_difference ≠ 0

t-value	P-Value
0.97	0.353

Wilcoxon test:

Sample	N	Median	CI for η	Achieved Confidence
Difference	12	-0.575	(-2.1, 0.8)	94.54%

Fig. 5.17 Minitab output for the paired t-test example and nonparametric alternative for samples from two sides of a tablet press

in a failure to reject the null hypothesis. The results did not prove that the results for the two chemists were the same or identical, simply that the study failed to find a difference.

5.5   One-Way Analysis of Variance (ANOVA)

The t-test was appropriate for one or two levels of the discrete independent variable, but what if there are three, four, or more discrete levels for the independent variable? The one-way *analysis of variance* provides an extension to k levels of the discrete independent variable. The analysis of variance is also referred to as the *F-test*, after Sir Ronald A. Fisher, a British statistician who developed this test during the 1920s (Salsburg, pp. 48–50). This section will focus on the one-way analysis of variance (abbreviated with the acronym *ANOVA*), which involves only one independent discrete variable and one dependent continuous variable.

The calculation involves an *analysis of variances* seen comparing the individual sample means for each level around a central grand mean for all the results. Like the t-test, the dependent variable represents data from a continuous distribution. The hypotheses being tested are:

$$H_0 : \quad \mu_1 = \mu_2 = \mu_3 \ldots = \mu_k$$
$$H_1 : \quad H_0 \text{ is false}$$

The null hypothesis states that there are no differences among the population means and that any fluctuations in the sample means are due to chance variability only. The ANOVA tests identify whether or not these differences are due to random chance or truly significant. One could write a decision rule for determining significance:

F-statistic:

$$F = \frac{MS_B}{MS_W}$$

Where: mean square within (MS_W):

$$MS_W = \frac{(n_1 - 1)S_1^2 + (n_2 - 1)S_2^2 + \cdots (n_k - 1)S_3^2}{N - K}$$

grand mean ($\bar{X}_G$)

$$\bar{X}_G = \frac{(n_1\bar{X}_1) + (n_2\bar{X}_2) + \cdots (n_k\bar{X}_k)}{N}$$

mean square between (MS_B):

$$MS_B = \frac{n_1(\bar{X}_1 - \bar{X}_G)^2 + n_2(\bar{X}_2 - \bar{X}_G)^2 + \cdots n_k(\bar{X}_k - \bar{X}_G)^2}{K - 1}$$

Fig. 5.18 Definitional formulas for a one-way analysis of variance

With α = ((researcher's choice))
reject H_0 if F-value is greater than ((absolute critical value Table B3)).

It is important to note that the alternative hypothesis does not state the $\mu_1 \neq \mu_2 \neq \cdots$ μ_k, only that a difference exists somewhere between at least two levels of the independent variable. They could all be different, but this cannot be determined with the ANOVA alone.

The results for a one-way ANOVA are presented as an F-ratio of the distances between the sample means divided by the dispersion within the sample means (Fig. 5.18). As the distances among the means become larger, the F-value will increase, and there is a greater likelihood for rejecting the null hypothesis. As the dispersion within the data increases, the denominator becomes larger, and the F-value will tend to decrease creating a greater likelihood of not rejecting the null hypothesis. The decision to reject or fail to reject the null hypothesis can be accomplished in one of two ways. First, identify a critical value of Table B3 in Appendix B based on a numerator degrees of freedom ($k - 1$ levels of the independent variable) and a denominator degrees of freedom ($N - k$ the total observations subtract the k levels of the independent variable). In Table B3 these $k - 1$ and $N - K$ degrees of freedom are noted as v_1 and v_2, respectively. If the calculated F-statistic exceeds the critical value, reject the null hypothesis. A second and simpler approach is to allow Minitab to calculate what is called an *ANOVA table* (Fig. 5.19). In the table Minitab labels the variability between the centers as "Factor" and within the variability as "Error." The F-statistic is the result of the calculation, and the reported p-value is the amount of potential error if the null hypothesis is rejected.

Notice in the ANOVA table in Fig. 5.19 that there is a column labeled *sum of squares*; this refers to a measurement called the sum of squares. As mentioned ear-

Traditional ANOVA Table:

Source	Degrees of Freedom	Sum of Squares[1]	Mean Square	F	$\underline{p}$
Between Groups	$k-1$	$III-II$	$\dfrac{III-II}{k-1}$	$\dfrac{MS_w}{MS_B}$	p-value[2]
Within Groups	$N-k$	$I-III$	$\dfrac{I-III}{N-k}$		
Total	$N-1$	$I-II$			

Example of an arbitrary Minitab report[3]:

Source	DF	Adj SS	Adj MS	F-Value	P-Value
Factor	2	50.13	25.07	1.88	0.165
Error	42	559.13	13.31		
Total	44	609.26			

[1] Sum of squares are intermediates derived from computational formulas (Figure 5.20)
[2] p-value is the amount of type I error based on the F-statistic
[3] Adjusted SS and MS represent minor corrections with minimal influence on results

Fig. 5.19 Analysis of Variance (ANOVA) Table

lier, there are two types of formulas. Definitional formulas display logic on how an equation works as seen in Fig. 5.18 where variability within the samples area defined (MS_W) and variability between centers is calculated (MS_B). But these *definitional formulas* would be difficult to write for computer code, so alternative computational formulas are available which are easier to use in computer code but give the exact same answer. The parallel *computational formulas* are seen in Fig. 5.20. This is where the measurements for the sum of squares are calculated (sum of squares between, sum of squares within, and sum of squares total). The resultant SS_B, SS_W, and SS_T entered into the ANOVA table. Divided by their respective degrees of freedom, they produce the mean squared values defined in Fig. 5.18.

To determine significance all that is needed is the ANOVA table, but Minitab provides additional supplemental information. These include a table with the means, standard deviations, and confidence intervals for each level of the independent variable and supplemental S, R-sq, R-sq(adj), and R-sq(pred). Their equations are presented in Fig. 5.21. The S is the pooled standard deviation measuring how far the sample data values differs from the ideal fitted values and is expressed in the same units as the dependent variable. It evaluates how well the ANOVA model describes the response. The smaller the S value, the better the model (in the case of the one-way ANOVA, fixed levels of the independent variable). But a small S value by itself

F-statistic:

$$F = \frac{MS_B}{MS_w}$$

Intermediate steps:

$$I = \sum_{k=1}^{K} \sum_{i=1}^{N} x_1^2 = (x_{a1})^2 + (x_{a2})^2 + \cdots (x_{kn})^2$$

$$II = \frac{\left[\sum_{k=1}^{K} \sum_{i=1}^{N} x_i\right]^2}{N} = \frac{\sum x_T^2}{N}$$

$$III = \sum_{k=1}^{K} \frac{\left[\sum_{i=1}^{N} x_i\right]}{N_k} = \frac{(\sum x_A)^2}{N_A} + \frac{(\sum x_B)^2}{N_B} + \cdots \frac{(\sum x_K)^2}{N_K}$$

Sums of squares:

$$SS_B = III - II$$
$$SS_w = I - III$$
$$SS_T = I - II$$

Mean squares:

$$MS = \frac{SS}{respective\ df}$$

Fig. 5.20 Computational formulas for a one-way analysis of variance

Pooled standard deviation:

$$S = \sqrt{SS_W}$$

Variation in response:

$$R^2 = \left(\frac{SS_B}{SS_T}\right) \times 100$$

Adjusted variation in response:

$$R_{adj}^2 = \left(1 - \frac{MS_w}{\frac{SS_T}{df_T}}\right) \times 100$$

Predictive variation in response:

$$r_{pred}^2 = 1 - \frac{PRESS}{SS_T}$$

Fig. 5.21 Additional measures supplemental to an ANOVA table

does not indicate that the model meets the model assumptions. One should also check the plot of the residual to verify the assumptions. The R^2 is the percentage of variation in the response that is explained by the fixed ANOVA model. The greater the R^2 value, the better the model fits the data. R^2 will always be between 0% and 100%. The SS_w, SS_B, and SS_T are all found in the ANOVA table. The adjusted R^2 (R_{adj}^2) is the percentage of the total variation that is explained by the model, but adjusted for more than one independent variable and based on the number of observations (N-way ANOVAs, Sect. 5.7). With the R_{adj}^2 it is possible to compare models that have different numbers of predictors (independent variables). R^2_{adj} will always increase when there are additional independent variables in the model, even when there is no real improvement to the model. When there are multiple predictor variables (independent variables), the R^2_{adj} can help with choosing the correct model. The final reported value with Minitab is the Predicted R^2 (R^2_{pred}) that is calculated using a formula that is equivalent to systematically remove each observation from the data set, estimate the regression equation (to be covered in Chap. 6), and determine how well the model predicts the removed observation. Calculations involve the PRESS statistic (predicted residual sum of squares) which can be used as a measure of predictive power. Its calculation is beyond the scope of this book but described by Allen (1974).

The one-way ANOVA is a parametric procedure; therefore it assumed that the populations from which the samples are taken are normally distributed and that there is homogeneity of variance among the sample dispersions.

5.5.1 An Extension of the Two-Sample t-test

If the one-way ANOVA is an extension of the two-sample t-test, then the same results should be seen if either test is run on data with only two levels of the discrete independent variable. For example, looking at the results in Fig. 5.12, it was determined using a two-sample t-test that there was a significant different between the two lots of drug ($t = -2.52$, $p = 0.016$). If this same sample data were evaluated using the one-way ANOVA, with one numerator degrees of freedom, similar results would be seen (Fig. 5.22). In this case for the t-test, equal variances are assumed, and the p-value increases slightly. Even though the t-statistic and F-statistic are different, the p-value for both tests is identical.

Fig. 5.22 Comparison of results for an one-way ANOVA with a two-sample t-test on the same sample data (Table 5.1)

Two-sample t-test, <u>assuming equal variances:</u>

t-value	DF	P-Value
-2.51	36	0.017

One-way ANOVA:

Source	DF	Adj SS	Adj MS	F-Value	P-Value
Factor	1	638857	638857	6.32	0.017
Error	36	3637501	101042		
Total	37	4276357			

5.5.2   Sample Size and Statistical Power

Like previous tests, sample size and statistical power may be important to the researcher in determining if there exist significant differences among the various levels of the independent variable. Minitab can perform these calculations, and the statistics for these determinations are discussed by Zar (pp. 189–195).

5.5.3   Multiple Comparisons (Post Hoc Tests)

Rejection of the null hypothesis in the one-way analysis of variance simply proves that some significant difference exists between at least two levels of the discrete independent variable. Unfortunately the ANOVA does not identify the exact location of the difference(s). *Multiple comparison* or *post hoc tests* can be used to reevaluate the data for a significant ANOVA and identify where the difference(s) exist while maintaining the overall type I error rate (α) at the same level as that is used to test the original null hypothesis for the one-way ANOVA. Assuming an analysis of variance was conducted with $\alpha = 0.05$ and the H_0 was rejected, then each multiple comparison tests will keep the error rate constant at 0.05.

A common error is to perform multiple t-tests between various two levels of the independent variable (called pair-wise combinations). However, by using multiple t-tests, the researcher actually compounds the type I error rate. This compounding of the error is referred to as the *experiment-wise error rate* and results in a much high type I error than 0.05. The calculation for the experiment-wise error rate is $1-(1-\alpha)^C$, where C is the number of pair-wise comparison that is possible. To avoid this Minitab has available four multiple comparison tests that can help identify the differences if they exist. Note that if the null hypothesis is not rejected, there is no reason for follow-up with a post hoc procedure. However, if the null hypothesis is rejected, selection of a multiple comparison is available by selecting tests under "Comparisons" in the ANOVA dialog box.

$$\text{Stats} \rightarrow \text{ANOVA} \rightarrow \text{One-way} \rightarrow \text{Comparisons}$$

The *Fisher's LSD test* (least significant difference test) and *Tukey's HSD test* (honestly significant difference test) are post hoc procedures that can be used for all pair-wise comparisons between two levels of the discrete independent variable. *Dunnett's test* is unique because it allows for one level of the independent variable to serve as a control and then each of the other levels compared to this control group (Dunnett 1955). *Hsu MCB test* (multiple comparison best test) compares each level of the independent variable to the level with either the largest or smallest mean (Hsu 1992). It is designed to identify levels that are the best, insignificantly different from the best, and those that are significantly different from the best. Equations for the first three tests are presented in Fig. 5.23.

Tukey's HSD Test[1]:

$$\mu_1 - \mu_2 = (\bar{X}_1 - \bar{X}_2) \pm q_{\alpha,k,N-k} \sqrt{\frac{MS_W}{n_1} + \frac{MS_W}{n_2}}$$

Fisher's LSD Test:

$$\mu_1 - \mu_2 = (\bar{X}_1 - \bar{X}_2) \pm t_{1-\alpha/2,N-k} \sqrt{\frac{MS_W}{n_1} + \frac{MS_W}{n_2}}$$

Dunnett's Test[2]:

$$\mu_{Control} - \mu_i = (\bar{X}_{Control} - \bar{X}_i) \pm q_{\alpha,p,N-k} \sqrt{MS_W \left(\frac{1}{n_{Control}} + \frac{1}{n_i}\right)}$$

[1] Percentage Point of the Studentized Range for q (Pearson and Hartley 1970)
[2] Critical Values of q for the Two-Tailed Dunnett's Test: (Dunnett)

Fig. 5.23 Equations for more commonly used post hoc comparisons

5.5.4   Nonparametric Alternatives

Much as the one-way ANOVA is an extension of the two-sample t-test, *Kruskal-Wallis test* is an equivalent nonparametric extension or generalization of the Mann-Whitney test for more than two levels of an independent discrete variable. As with other nonparametric procedures, it involves ranking the data across the multiple levels of the independent variable (Kruskal and Wallis 1952). The null hypothesis is actually stating that all of the population medians are equal (H_0: $\eta_1 = \eta_2 = \eta_3 \ldots = \eta_k$). The alternative hypothesis is that a difference exists somewhere among the population medians (not all η's are equal) or more simply stated:

$$H_0 : \text{Samples are from the same population}$$

$$H_1 : \text{Samples are drawn from different populations}$$

The determination of significance is a *p*-value based on the conversion of an *H*-score calculated from the sum of the ranks for each level of the independent variable. There is a secondary *H*-score reported with Minitab that is more conservative and adjusts for ties in ranking across the various levels of the independent variable. The equations for calculating the Kruskal-Wallis test (with and without a correction for tied ranks) are discussed by Daniels (1978) (pp. 200–206).

   Mood's test is similar to Kruskal-Wallis in that it compares the median values for two or more levels of a discrete independent variable. It calculates a range of values likely to include the difference among population medians (Mood 1950). The determination of significance is based on a 2-by-*k* chi square test of independence (Chap. 6) where the number of observations above the median and below the median is determined for each discrete level and compared to all the other levels of the independent variable. The formula for Mood's test is presented in Hollander and Wolfe (1999) (pp. 189–193).

In choosing which one of the two tests to use, Kruskal-Wallis is recommended because it is considered the more powerful of the two tests and the associated p-value for the correction of ties is more appropriate.

5.5.5 Minitab Applications

One-way analysis of variance and nonparametric alternatives is appropriate when there are two or more levels of a discrete independent variable and a continuous dependent variable. One-way ANOVAs with their corresponding sample size determination and power can be calculated with Minitab, as well as Kruskal-Wallis and Mood's tests. Steps for performing these tests are presented below.

5.5.5.1 One-Way ANOVA

Procedure	Stats → ANOVA → One-way
Data input	Type selected on dialog box: (1) data with each column a variable on the worksheet ("Response" is the dependent variable and "Factor" is the independent variable); or (2) each level of the independent variable is on a different column in the worksheet.
Options	Automatic default for a 95% confidence interval for a two-tailed test. These can be changed to a different degree of type I error if required or a one-tailed test is desired.
Graphs	By default, confidence intervals will be created and plotted for each level of the independent variable (interval plot). This option can be turned off. Other graphic options include an individual value plot (dot plot) or box plot for each level of the independent variable. A variety of residual plots are also available (differences between the data points and means).
Comparisons	Available post hoc procedures that can be selected (see Multiple Comparisons below).
Report	The most important part of the report is the ANOVA table with its F-statistic and corresponding p-value. Also reports the sample means and standard deviations for all levels of the independent variable and some model summary results (S representing the pooled standard deviation for the data set).
Interpretation	If the p-value in the ANOVA table is less than 0.05, reject the null hypothesis in favor of the alternative hypothesis that some significant difference exists among the levels of the independent variable.

5.5.5.2 Multiple Comparisons

If there is a significant one-way ANOVA, the procedure can be rerun to identify where the significant difference(s) exist. If the one-way ANOVA is not significant ($p > 0.05$), there is no reason to do any post hoc multiple comparisons.

Procedure	Stats → ANOVA → One-way → Comparisons
Data input, options, and graphs	Leave the same as the original one-way ANOVA.
Comparisons	Post hoc procedures to be used can be selected (Tukey, Fisher, Dunnett, and/or Hsu) as well as the output reports (interval plot of differences, grouping information or tests). Tests are the most relevant because it will create confidence intervals with associated p-values.
Report	The original information for the one-way ANOVA will be presented again, followed by the multiple comparison information. The grouping report couples nonsignificant levels together. The interval plot graphic shows the relationship of the different paired sets which are confidence intervals. The test report presents all the paired comparisons with their individual statistic and associated p-value.
Interpretation	Using the individual test results, any p-value less than 0.05 would be considered significant. Visual inspection of the confidence interval graphic would identify significance as any interval where zero is not a possible outcome.

5.5.5.3 Sample Size and Statistical Power

Procedure	Stat → Power and Sample Size → One-way ANOVA
Data input	Standard deviation is required and two of the following: (1) sample size, (2) detectable difference, and (3) power. Program will solve for the third item not entered. Multiple values can be selected for any one of the two input items (e.g., sample sizes of 10, 20, 30, and 40).
Options	Automatic default for a 95% confidence interval for a two-tailed test. These can be changed to a different degree of type I error if required or a one-tailed test is desired.
Graphs	By default will produce a power curve, which can be turned off under "Graph." Multiple sample sizes can be added to create a graph with multiple curves and power points.

| Report | Results for the third item not entered (e.g., if sample size and detectable difference are entered, results will report the power). Multiple entries (e.g., sample size) will report results for each sample size entered. |
| Interpretation | Helps identify the power for different sample sizes or different detectable differences or the required sample size for a desired power and detectable difference. |

5.5.5.4   Nonparametric Alternative: Kruskal-Wallis Test

Procedure	Stats → Nonparametric → Kruskal-Wallis
Data input	The dependent variable column is selected for "Response" and the independent variable column for "Factor."
Results	Kruskal-Wallis H-statistic and associated p-value are reported for both corrections for ties and without the correction. A median for each level of the independent variable is reported with associated Z-values.
Interpretation	If the p-value is less than 0.05, reject the null hypothesis in favor of the alternative hypothesis that there is a significant difference among the levels of the independent variable. The Z-values in the reports for medians might be useful in identifying post hoc differences. For levels with a Z-value greater than 1.96 (95% confidence) are significantly different than the other levels in the study.

5.5.5.5   Nonparametric Alternative: Mood's Median Test

Procedure	Stats → Nonparametric → Mood's Median Test
Data input	The dependent variable is selected for "Response" and the independent variable for "Factor." If "Store residuals" or "Store fits" are checked, these results will appear in the next available columns on the worksheet.
Results	The results for a chi square statistic (Chap. 6) will be calculated and a corresponding p-value.
Interpretation	If the p-value is less than 0.05, reject the null hypothesis in favor of the alternative hypothesis that there is a significant difference among the levels of the independent variable.

5.5.6   Examples

The quality control laboratory at pharmaceutical company wants to purchase new dissolution equipment (testers). Five different testers are evaluated to determine if there is significant difference in their results based on testing of a single product. Tablets are selected from the same batch of a drug and randomly divided into five groups for testing on the various dissolution testers. Data from three individual runs were combined for each tester with some testers having six vessels and others eight vessels (Table 5.5). Is there any significant difference in the dissolution results (at 15 minutes) based on the testers studied?

The most appropriate test would be a one-way ANOVA since there are five discrete levels of the independent variable and a continuous dependent variable (percent). The results in the ANOVA table indicate that there is a significant difference somewhere among the five testers with $F = 8.89$ and $p < 0.001$ (Fig. 5.24). So, the null hypothesis that all the testers are equal at 15 minutes is rejected, but the ANOVA

Table 5.5   Dissolution results (percent) at 15 minutes for five different apparatuses

	Tester A	Tester B	Tester C	Tester D	Tester E
	83.7	77.8	78.8	77.8	75.3
	76.8	77.4	78.4	77.4	76.1
	80.7	77.3	78.3	77.3	76.0
	80.1	78.8	78.2	76.2	69.9
	80.5	77.8	79.3	78.8	77.5
	79.9	76.4	76.4	74.6	73.3
	80.3	78.8	81.1	79.3	72.0
	81.1	75.3	76.6	77.2	77.7
	80.7	76.8	79.3	71.9	80.2
	81.6	75.9	77.4	83.5	75.1
	82.3	76.9	83.2	75.2	73.5
	78.6	74.4	74.8	79.6	71.5
	73.9	72.3	79.2	77.8	74.6
	84.5	76.9	78.5	79.0	73.2
	79.1	79.1	79.3	77.5	77.0
	79.5	80.2	73.7	80.1	76.9
	84.5	81.6	78.2	74.5	77.4
	82.6	82.7	83.7	82.7	80.7
			78.1	77.1	
			79.3	78.3	
			79.6	78.6	
			77.2	76.2	
			74.8	73.8	
			73.9	72.9	
Mean =	80.58	77.58	78.22	77.39	75.44
Standard deviation =	2.61	2.47	2.48	2.76	2.88
n =	18	18	24	24	18

One-way ANOVA:

Source	DF	Adj SS	Adj MS	F-Value	P-Value
Factor	4	248.3	62.081	8.89	0.000
Error	97	677.5	6.985		
Total	101	925.9			

Tester	N	Mean	StDev	95% CI
A	18	80.578	2.610	(79.341, 81.814)
B	18	77.578	2.469	(76.341, 78.814)
C	24	78.221	2.478	(77.150, 79.292)
D	24	77.387	2.764	(76.317, 78.458)
E	18	75.439	2.881	(74.203, 76.675)

Pooled StDev = 2.64289

Graphic results:

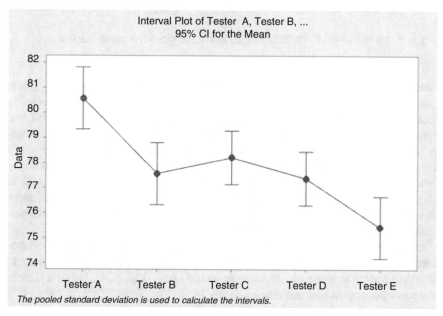

Fig. 5.24 Minitab output and graphics for the results for one-way ANOVA for five dissolution testers

Difference of Levels	Difference of Means	SE of Difference	95% CI	t-value	Adjusted P-Value
Tester B - Tester A	-3.000	0.881	(-5.448, -0.552)	-3.41	0.008
Tester C - Tester A	-2.357	0.824	(-4.647, -0.067)	-2.86	0.041
Tester D - Tester A	-3.190	0.824	(-5.480, -0.900)	-3.87	0.002
Tester E - Tester A	-5.139	0.881	(-7.587, -2.691)	-5.83	0.000
Tester C - Tester B	0.643	0.824	(-1.647, 2.933)	0.78	0.936
Tester D - Tester B	-0.190	0.824	(-2.480, 2.100)	-0.23	0.999
Tester E - Tester B	-2.139	0.881	(-4.587, 0.309)	-2.43	0.117
Tester D - Tester C	-0.833	0.763	(-2.953, 1.287)	-1.09	0.810
Tester E - Tester C	-2.782	0.824	(-5.072, -0.492)	-3.38	0.009
Tester E - Tester D	-1.949	0.824	(-4.239, 0.341)	-2.36	0.134

Fig. 5.25 Minitab output for the Tukey's HSD results comparing five dissolution testers

does not identify where difference(s) exist. Visually, looking at the default graphic generated by Minitab in Fig. 5.24, it appears that Apparatuses A and E are probably different from each three other testers, but are there other differences? Post hoc procedures are performed using both the Tukey's HSD test and Fisher's LSD test (Figs. 5.25 and 5.26). Similar results are seen with significant differences ($p < 0.05$) observed between Tester A and all the other testers and between Tester E and all the other testers. No significant difference was identified between Tests B, C, and D. In this example it appears that Tukey's HSD test is a little more conservative and harder to reject similarities based on larger p-values, with five nonsignificant results compared to only three for Fisher's LSD test.

Assume that Tester C is the dissolution apparatus currently available in the laboratory and the researcher's original intent was to compare the other four testers to this "control." The same ANOVA results would occur (Fig. 5.24), but in this case, the appropriate post hoc test procedure would be to do Dunnett's tests where C is compared to the other four testers (Fig. 5.27). In this case differences were not identified with Testers B and D, but there was a significant difference between the control and Tester A ($p = 0.019$) and Tester E ($p = 0.004$).

If nonparametric tests were performed on the same data, similar results would be seen for both Kruskal-Wallis and Mood's test (Fig. 5.28). Results are identical to the ANOVA with all three tests having a $p < 0.001$. As noted in Sect. 4.2.3, Minitab truncates very small p-values and reports the results as 0.000. Since it is impossible to

Difference of Levels	Difference of Means	SE of Difference	95% CI	t-value	Adjusted P-Value
Tester B - Tester A	-3.000	0.881	(-4.748, -1.252)	-3.41	0.001
Tester C - Tester A	-2.357	0.824	(-3.992, -0.721)	-2.86	0.005
Tester D - Tester A	-3.190	0.824	(-4.826, -1.555)	-3.87	0.000
Tester E - Tester A	-5.139	0.881	(-6.887, -3.390)	-5.83	0.000
Tester C - Tester B	0.643	0.824	(-0.992, 2.279)	0.78	0.437
Tester D - Tester B	-0.190	0.824	(-1.826, 1.445)	-0.23	0.818
Tester E - Tester B	-2.139	0.881	(-3.887, -0.390)	-2.43	0.017
Tester D - Tester C	-0.833	0.763	(-2.348, 0.681)	-1.09	0.277
Tester E - Tester C	-2.782	0.824	(-4.417, -1.146)	-3.38	0.001
Tester E - Tester D	-1.949	0.824	(-3.584, -0.313)	-2.36	0.020

Fig. 5.26 Minitab output for the Fisher's LSD test results comparing five dissolution testers

Difference of Levels	Difference of Means	SE of Difference	95% CI	t-value	Adjusted P-Value
Tester A - Tester C	2.357	0.824	(0.301, 4.413)	2.86	0.019
Tester B - Tester C	-0.643	0.824	(-2.699, 1.413)	0.78	0.861
Tester D - Tester C	-0.833	0.763	(-2.737, 1.070)	1.09	0.663
Tester E - Tester C	-2.782	0.824	(-4.838, -0.726)	3.38	0.004

Fig. 5.27 Minitab output for the Dunnett's results comparing four dissolution testers with Tester C serving as the control

have no type I error and 100% confident with an inferential statistics, it is advisable to report such results as $p < 0.001$ since no information is available after the third zero.

There may be some concern over the statistical power with this small data set. In this case the pooled standard deviation is reported in Minitab as the 2.64, and the n is 18 because it is the smaller sample sizes. Assume the desired detectable difference of 5%; which in this case is the 5% of the weighted grand mean for 15 minutes (0.05 × 77.83 = 3.89). Under these conditions the statistical power for this test was 0.949 (Fig. 5.29). Only a sample size of 12 results per tester would be needed to meet a goal of 80% power to detect a 5% difference.

Kruskal-Wallis Test:

Null hypothesis	H_0: All medians are equal		
Alternative hypothesis	H_1: At least one median is different		
Method	**DF**	**H-Value**	**P-Value**
Not adjusted for ties	4	27.50	0.000
Adjusted for ties	4	27.51	0.000

Mood's Test:

Null hypothesis	H_0: The population medians are all equal	
Alternative hypothesis	H_1: The population medians are not all equal	
DF	**Chi-Square**	**P-Value**
4	27.83	0.000

Fig. 5.28 Minitab output for the nonparametric alternative results for five dissolution testers

Power given 18 observations to detect a 5% difference :

Maximum Difference	Sample Size	Power
3.89	18	0.948921

The sample size is for each level.

Required observations for power of 0.80 to detect a 5% difference:

Maximum Difference	Sample Size	Target Power	Actual Power
3.89	12	0.8	0.800725

The sample size is for each level.

Fig. 5.29 Minitab output for sample size and power determinations from the one-way ANOVA example for dissolution testers

In a second example, Acme Chemical and Dye received three batches of an oil from the same raw material supplier shipped to three different production sites. Samples were drawn from drums at each location and compared to determine if the viscosity was the same for each batch (Table 5.6). Are the viscosities the same regardless of the batch? The results of the Minitab analysis are presented in Fig. 5.30 and indicate that there is no significant difference between the three batches of oil ($F = 1.99, p = 179$). Since no significant differences were identified, there is no need to perform any post hoc tests. It is important to note that Acme did not prove that the

Table 5.6 Viscosity of different batches of a product

	Batch A	Batch B	Batch C
	10.23	10.24	10.25
	10.22	10.28	10.24
	10.28	10.20	10.21
	10.27	10.21	10.18
	10.30	10.26	10.22
Mean =	10.26	10.238	10.22
S.D. =	0.034	0.033	0.027
n =	5	5	5

One-way ANOVA:

Source	DF	Adj SS	Adj MS	F-Value	P-Value
Factor	2	0.004013	0.002007	1.99	0.179
Error	12	0.012080	0.001007		
Total	14	0.016093			

Graphic results:

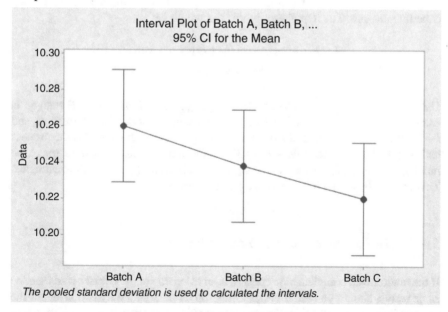

The pooled standard deviation is used to calculated the intervals.

Fig. 5.30 Minitab output and graphic results for one-way ANOVA comparing viscosity for three batches of oil

three batches were the same (null hypothesis) but simply failed to find a difference. Test for equivalence will be presented in Chap. 7.

5.6   Randomized Complete Block Design

The paired t-test also can be expanded to more than two points of comparison (e.g., results for the same batch tested by three or more different methods). The test is one of the several randomized block designs developed in the 1920s by Sir Ronald Fisher to evaluate methods for improving agricultural experiments (Salsburg, 2002 pp. 50–51). As part of his agricultural experiments and to eliminate variability between different locations of fields, Fisher's research design first divided the land into *blocks*. The area within each block was assumed to be relatively homogeneous. Then each of the blocks was further subdivided into *plots*, and each plot within a given block received one of the treatments under consideration. Therefore, only one plot within each block received a specific treatment, and each block contained plots that represented all the different treatments.

Expanding on the previous example of the paired t-test (Sect. 5.4.4) where two methods were compared, what if three, four, or k methods were evaluated? In this case each of the ten batches (counted as j batches) would be considered a block, and the single response for each treatment level (k methods) would consist of $k \times j$ experimental unit. The hypotheses in this test would be:

$$H_0 : \text{No difference in the treatment levels}$$
$$H_1 : \text{A difference exists in the treatment levels}$$

Using mathematical manipulations (De Muth, pp. 216–219), a new F-statistic is calculated in an ANOVA table based on $k - 1$ numerator degrees of freedom and $j - 1$ denominator degrees of freedom. Once again, Minitab simplifies the process and reports the ANOVA table with an F-statistic and p-value. If $p < 0.05$, reject the null hypothesis, and find that there is significant difference(s) among the treatments. This will be illustrated in an example presented below.

5.6.1   An Extension of the Paired t-test

If the randomized complete block design is an extension of the paired t-test, then the same results should be seen if either test is run on data with only two levels of the discrete independent variable. For example, looking at the results in Fig. 5.15, it was determined using a paired t-test that there was no significant difference between the two methods ($t = 1.29, p = 0.229$). If this same sample data were evaluated using the randomized complete block design, with one numerator degree of freedom, similar results would be seen (Fig. 5.31). Even though the F-value is different from the t-value, the p-value for both tests is identical.

Paired t-test, assuming equal variances:

t-value	P-Value
1.29	0.229

Randomized complete block design ANOVA:

Source	DF	Adj SS	Adj MS	F-Value	P-Value
Batch	9	5.8505	0.6501	6.00	0.007
Method	1	0.1805	0.1805	1.67	0.229
Error	9	0.9745	0.1083		
Total	19	7.0055			

Fig. 5.31 Comparison of results for a randomized complete block design results with paired t-test on the same data (Table 5.3)

5.6.2 Nonparametric Alternative

The Friedman procedure is available in Minitab and can be employed for data meeting the design for a randomized block design, but that fails to conform to the criteria for parametric procedures (normality and homogeneity). This randomized block design can be considered a nonparametric extension of the Wilcoxon matched pairs test to more than two treatment levels or times. The null hypothesis is that the treatment has no effect (similar to the complete randomized block design). The equation for calculating Friedman test is discussed by Conover 1999 (pp. 369–371).

5.6.3 Minitab Applications

This test is an extension of the paired t-tests to two or more levels of an independent variable that are paired and a continuous dependent variable. The nonparametric alternative is the Friedman test. Steps for performing these tests are presented below.

5.6.3.1 Randomized Complete Block Design

Procedure	Stats → ANOVA → General Linear Model → Fit General Linear Model
Data input	The dependent variable is selected for "Responses," the independent variable and blocking columns for "Factors."
Options	Automatic default for a 95% confidence interval for a two-tailed test. These can be changed to a different degree of type I error if required or a one-tailed test is desired.

Random/nest, model, coding, stepwise, graphs, and storage	Leave default values (beyond the scope of this book).
Results	A variety of information is presented in the report. Most can be turned off and just refer to the "Analysis of Variance."
Report	The most important part of any report is the ANOVA table with its F-statistic and associated p-value for both the main treatment effect and the blocking effect.
Interpretation	If the p-value in the ANOVA table is less than 0.05, reject the null hypothesis in favor of the alternative hypothesis that some significant difference exists among the levels of either or both the treatment factor or the blocking factor.

5.6.3.2   Nonparametric Alternative: Friedman Test

Procedure	Stats → Nonparametric → Friedman
Data input	The dependent variable is selected for "Response," the independent variable for "Factor," and the blocking column for "Blocks." If "Store residuals" or "Store fits" are checked, these results will appear in the next available columns on the worksheet.
Results	The results for a chi square statistic (Chap. 6) will be calculated and a corresponding p-value (both adjusting for ties and not adjusting).
Interpretation	If the p-value is less than 0.05, reject the null hypothesis in favor of the alternative hypothesis that there is a significant difference among the levels of the independent variable.

5.6.4   Example

Using the same example as the paired t-test (Sect. 5.4.4 and Table 5.3), assume instead comparing only two methods, there are now three methods being tested. The results for Method C are 17.6, 17.7, 19.2, 17.8, 18.6, 17.6, 17.6, 18.1, 18.2, and 17.7 for batches one through ten, respectively. A complete randomized block design can be used to the compare the three methods. The result (Fig. 5.32) shows no significant difference between the three methods used to analyze the batches ($F = 1.69$, $p = 0.212$) even though there was a significant difference between the ten batches being tested ($F = 16.15$, $p < 0.001$).

As seen in the results for the Friedman test, all three methods give similar medians and sums of the rank scores. Note that the Friedman test only evaluates the treatment

Complete randomized block design:

Source	DF	Adj SS	Adj MS	F-Value	P-Value
Batch	9	8.3413	0.92681	16.15	0.000
Method	2	0.1940	0.09700	1.69	0.212
Error	18	1.0327	0.05737		
Total	29	9.5680			

Friedman test:

Method	N	Median	Sum of Ranks
A	10	17.6667	18.0
B	10	17.7833	21.5
C	10	17.7500	20.5
Overall	30	17.7333	

Null hypothesis H_0: All treatment effects are zero
Alternative hypothesis H_1: Not all treatment effects are zero

Method	DF	Chi-Square	P-Value
Not adjusted for ties	2	0.65	0.723
Adjusted for ties	2	0.74	0.690

Fig. 5.32 Minitab output for the complete randomized block example and nonparametric alternative

effect (methods) and not the blocking effect. Results are similar with both methods failing to reject the null hypothesis of no difference between the methods. But the Friedman produces a substantially larger *p*-value that is its parametric partner.

5.7   *N*-Way ANOVAs

N-Way ANOVAs allow the researcher to compare multiple independent variables (called predictor variables) and their effects on one dependent variable (response variable). These tests are extensions of the one-way ANOVA and determine the significance of each independent variable as well as potential interactions or confounding with other independent variables. The simplest is the two-way ANOVA.

5.7.1   *Two-Way Analysis of Variance*

In a two-way ANOVA, the researcher is interested in the major effects of two discrete independent variables (factors) on the dependent variable (response) and their potential interaction. Sometime refer to as factorial designs, the number of factors (synonym for variables) and levels within each factor determines the dimensions of this design. For example, if one variable consists of three levels and second only

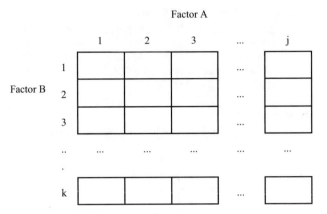

Fig. 5.33 Layout design for a two-way analysis of variance

two, it would be presented as a 3 × 2 (reads as a three-by-two) factorial design. In contrast to the complete randomized block design, in these factorial designs, there must be more than one observation per cell. The design for a two-way analysis of variance is presented in Fig. 5.33. The term *cell* refers to information within the space represented by one column and one row. If there were only one observation per cell, there would be no variance within the cells; therefore, calculation of a required sum-of-squares error term would be impossible. Each of the two major factors is a discrete independent variable, and the significance of each factor is measured based on a single continuous dependent variable.

In a two-way ANOVA, three hypotheses are tested simultaneously: two testing for the main effects and one for the interaction:

$$H_{01} : \mu_{A1} = \mu_{A2} = \mu_{A3} = \dots \mu_{AJ} \qquad \text{(Main effect of Factor A)}$$
$$H_{02} : \mu_{B1} = \mu_{B2} = \mu_{B3} = \dots \mu_{BK} \qquad \text{(Main effect of Factor B)}$$
$$H_{03} : \left(\mu_{A1,B1} - \mu_{A1,B2} \right) = \left(\mu_{A2,B1} - \mu_{A2,B2} \right) = \text{etc.} \quad \text{(Interaction of A and B)}$$

At the same time, there are three mutually exclusive and exhaustive alternative hypotheses to complement each of the null hypotheses:

$$H_{11} : \ H_{01} \text{ is false}$$
$$H_{12} : \ H_{02} \text{ is false}$$
$$H_{13} : \ H_{03} \text{ is false}$$

Because the two-way analysis of variance deals with two independent variables, there may be different critical F-values (F_c) depending on the size of the ANOVA matrix. Taken from Table B3 in Appendix B, F_c is associated with each null hypothesis being tested, and these are directly associated with the number of rows and columns presented in the design matrix. The symbols used are j for the number of levels of the column variable, k for the number of levels of the row variable, N_k for the total number of observations, and n for the number of observations per cell in the case of equal cell sizes.

Complicated mathematical equation is used to create a new ANOVA table (De Muth, pp. 272–285). The evaluation involves $j - 1$ degrees of freedom for one factor, $k - 1$ degrees of freedom for the second factor, $(k - 1) \times (j - 1)$ degrees of freedom for the interaction, and $k \times j \times (n - 1)$ degrees of freedom for the error term and results in an ANOVA table. Fortunately Minitab greatly simplifies the process, does all the mathematical manipulation on the data, and produces an ANOVA table. The result is in the table with three separate F-values and associated p-values (representing each of the two factors and their interaction). Determination of significance would be rows with reported p-values less than 0.05.

5.7.1.1 Minitab Application

The steps for the Minitab execution of a two-way analysis of variance when there is a continuous dependent variable and two discrete independent variables (factors) are as follows:

Procedure	Stats → ANOVA → General Linear Model → Fit General Linear Model
Data input	Selected the dependent variable ("Response") and the independent variables ("Factors").
Random/nest, coding, stepwise, and storage	Leave default values (beyond the scope of this book).
Models	Highlight the names of the two independent variables and "2" for "Interactions through order." What should appear under model are each factor and factor 1 × factor 2.
Options	Automatic default for a 95% confidence interval for a two-tailed test. These can be changed to a different degree of type I error if required or a one-tailed test is desired.
Graphs	A variety of residual plots are available.
Results	A variety of information is presented in the report. Most can be turned off and just refer to the "Analysis of Variance."
Report	The most important part of the report is the ANOVA table with its F-statistics and associated p-values.
Interpretation	For those p-values in the ANOVA table that are less than 0.05, reject the null hypothesis in favor of the alternative hypothesis that some significant difference exists among the levels of the independent variables or the interaction being tested.

5.7.1.2   Examples

Two technicians are assigned to do analyses using two different methods. Samples from a single batch of a product are randomly assigned to the different test conditions and different analysts (Table 5.7). The research questions or alternative hypotheses are the following: (1) is there a significant difference in the performance of the two analysts; (2) is there a significant difference between the two methods; and (3) is there any interaction between the analyst and method tested? The hypotheses being tested would be:

$$H_{01} : \mu_{\text{Analyst A}} = \mu_{\text{Analyst B}}$$
$$H_{02} : \mu_{\text{Method 1}} = \mu_{\text{Method 2}}$$
$$H_{03} : \left(\mu_{\text{Anaylst A,Day 1}} - \mu_{\text{Formula B,Day 1}} \right) = \left(\mu_{\text{Analust A,Day 2}} - \mu_{\text{Analyst B,Day 2}} \right)$$

$$H_{11} : H_{01} \text{ is false}$$
$$H_{12} : H_{02} \text{ is false}$$
$$H_{13} : H_{03} \text{ is false}$$

The results of the Minitab analysis are presented in Fig. 5.34. Based on the analysis, there was no significant difference based on the method used ($F = 0.42$, $p = 0.526$), but there were significant differences based on the analyst ($F = 14.09$, $p = 0.001$) and a much larger significant interaction ($F = 43.15$, $p < 0.001$). The interaction becomes obvious when observing the means for each analyst/method as seen in Fig. 5.35. Analyst B reports larger results with Method 1, Analyst B reports smaller results with Method 1, and these results reverse when using Method 2.

Table 5.7 Comparison of results for different analysts using different methods of analysis

Results for % label claim		
	Analyst A	Analyst B
	98.5	100.3
	100.1	100.5
Method 1	98.9	100.0
	99.1	98.9
	99.5	99.5
	98.7	100.4
	101.6	99.5
	101.2	98.6
Method 2	102.3	98.5
	100.8	97.8
	100.5	98.2
	100.7	96.9

Source	DF	Adj SS	Adj MS	F-Value	P-Value
Method	1	0.2017	0.2017	0.42	0.526
Analyst	1	6.8267	6.8267	14.09	0.001
Method*Analyst	1	20.9067	20.9067	43.15	0.000
Error	20	9.6900	0.4845		
Total	23	37.6250			

Fig. 5.34 Minitab output for a two-way ANOVA comparing different analysts and different methods

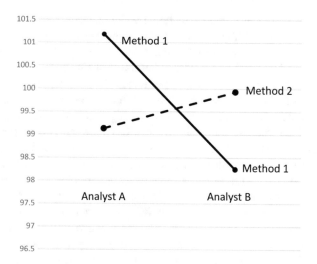

Fig. 5.35 Graphic comparison of outcomes for different analysts and different methods

Source	DF	Adj SS	Adj MS	F-Value	P-Value
Speed	3	0.365	0.1215	0.08	0.973
Filler	2	33.250	16.6250	10.35	0.000
Speed*Filler	6	2.417	0.4028	0.25	0.958
Error	84	134.875	1.6057		
Total	95	170.906			

Fig. 5.36 Minitab output for the two-way ANOVA of different tablet press speeds and different excipients

In a second example, a formulator is testing three different excipients for a tablet (lactose, microcrystalline cellulose, and dicalcium phosphate) and will run the tablet press at different speeds (De Muth, p. 309). The outcome of interest is tablet hardness (kP). In the setup of the two-way ANOVA, the "Response" was hardness, the two "Factors" were excipient and speed, and there was an "Excipient × Speed" added under the "Model" option for the interaction of the two independent variables. The results of the study are presented in Fig. 5.36, and it was found that there was no significant difference based on the speed of the tablet press and no significant interaction between the excipient and press speed. However, there was a significant difference in the tablet hardness based on the excipient used in the study ($F = 10.35$, $p < 0.001$). Therefore, the formulator should use the excipient best for the appropriate tablet hardness and not worry about the speed of the tablet press.

5.7.2   Three-Way and Larger N-Way Analysis of Variances

With the two-way ANOVA, there were only two independent variables and only one potential interaction affecting the outcomes with the dependent variable. The advantage of these multifactor designs is the increased efficiency for comparing different levels of several independent variables (factors) in a single study, instead of conducting several separate single-factor experiments and possibly missing interactions between variables. However, as the number of independent variables increases, the number of possible outcomes increases, and designs get extremely complicated to perform and interpret, especially the interactions between two or possibly more variables. In the case of the three-way ANOVA, there are three independent variables, three two-way interactions, and one three-way interaction to evaluate. For a four-way ANOVA, it increases to four independent variables, six two-way interactions, three three-way interactions, and one four-way interaction possible. Imaging the outcomes with five-way or six-way ANOVAs! As will be seen in the following example, the evaluation of the results involves a much expanded ANOVA table where each variable and interaction is represented by a separate row with an F-statistic and associated p-value. Interpretation of these results would be to identify factors and interactions that are significant, usually where p-values are less than 0.05.

N-Way ANOVAs are also referred to as MANOVA or multivariate analysis of variance. MANOVAs are intended for large research studies where there are a number of different variables to be assessed. Multiple one-way ANOVAs can result in a compounding of the error rate when the same data is used repeatedly (similar to the multiple comparisons discussed in Sect. 5.5.3). MANOVAs can detect mean differences for a number of different groups and their potential interactions where the type I error rate remains constant.

Well beyond the scope of this book, there are ANCOVAs (*analysis of covariance*) and MANCOVA (*multivariate analysis of covariance*), which is a combination of ANCOVA and MANOVA designs. The MANCOVA is used when the researcher wishes to detect mean differences among a number of different levels of the independent variable while holding one or more other variables constant. The MANCOVA is useful when a variety of levels is evaluated on a number of different measures. Also, there are fractional factorial designs or incomplete block designs. These are experimental designs in which only some of the treatment blocks are included in the statistical analysis. Because of this increased complexity, factorial designs involving more than three factors pose difficulties in the interpretation of the interaction effects. Therefore, most factorial designs are usually limited to three factors. Minitab can handle several of these designs. Excellent references for these more complex designs are described by Petersen (1985) and Box et al. (1978).

To simplify the process and make interpretation of the results easier, try to use the design of experiments to reduce the number of independent variables that will be tested in any one statistical evaluation. Minitab can handle this complex type of data using the same procedure as the two-way ANOVA.

5.7.2.1   Minitab Applications

The steps for applying a *N*-way ANOVA are similar to those for the two-way ANOVA test and are as follows:

Procedure	Stats → ANOVA → General Linear Model → Fit General Linear Model
Data input	Select the dependent variable ("Response") and the independent variables ("Factors").
Random/nest, coding, stepwise, and storage	Leave default values (beyond the scope of this book).
Models	Similar to the two-way ANOVA, the only difference is to include every *N*-way interaction under "Interactions through order." For example if there are factors A, B, and C, the model should include A × B, A × C, B × C, and A × B × C.
Options	Automatic default for a 95% confidence interval for a two-tailed test. These can be changed to a different degree of type I error if required or a one-tailed test is desired.
Graphs	A variety of residual plots are available.
Results	A variety of information is presented in the report. Most can be turned off and just refer to the "Analysis of Variance."
Report	The most important part of the report is the ANOVA table with its *F*-statistics and associated *p*-values.
Interpretation	For those *p*-values in the ANOVA table less than 0.05, reject the null hypothesis in favor of the alternative hypothesis that some significant difference exists among the levels of the independent variables or the interaction being tested.

Source	DF	Adj SS	Adj MS	F-Value	P-Value
Lot	3	3.24	1.0786	0.26	0.852
Technician	2	37.95	18.9771	4.62	0.011
Equipment	1	3.85	3.8461	0.94	0.334
Temperature	1	1.01	1.0112	0.25	0.620
Lot*Technician	6	240.52	40.0873	9.76	0.000
Lot*Equipment	3	28.68	9.5599	2.33	0.076
Lot*Temperature	3	4.19	1.3976	0.34	0.796
Technician*Equipment	2	58.11	29.0560	7.07	0.001
Technician*Temperature	2	0.35	0.1744	0.04	0.958
Equipment*Temperature	1	0.13	0.1301	0.03	0.859
Lot*Technician*Equipment	6	26.71	4.4518	1.08	0.374
Lot*Technician*Temperature	6	8.99	1.4975	0.36	0.901
Lot*Equipment*Temperature	3	1.93	0.6419	0.16	0.926
Technician*Equipment*Temperature	2	0.01	0.0031	0.00	0.999
Lot*Technician*Equipment*Temperature	6	4.14	0.6906	0.17	0.985
Error	192	788.72	4.1079		
Total	239	1208.32			

Fig. 5.37 Minitab output for a *N*-way ANOVA

5.7.2.2 Example

As an example, Fig. 5.37 shows the results of a study looking at the percent recovered in an analysis considering different lots of the product, different technicians, different equipment, and different temperatures. As noted by the degrees of freedom $(n-1)$, there were four drug lots tested by three technicians on two different pieces of equipment and at two different temperatures. The results show that the technician appears to be the most important sole factor ($p = 0.011$) with significant interactions with the drug lot ($p < 0.001$) and the equipment ($p = 0.001$). No other significant difference was noted because all the other p-values were greater than 0.05. Notice that this figure is nothing more than an expansion of the one-way ANOVA table in Fig. 5.19 with many more factors and interactions with their associated p-value, but the interpretation is the same.

5.8 Z-Tests of Proportions

All the previous tests in this chapter have dealt with discrete independent variable(s) and a continuous dependent variable. In contrast the Z-tests of proportions evaluate differences in percent (expressed as proportions) outcome results for one or two levels of a discrete independent variable. For example, analysis for samples from batch A has a failure rate of 1.5%, whereas batch B has a failure rate of 4.0%. Do the proportions 0.015 and 0.040 represent a statistically significant difference or is the difference due to chance error? To answer the question, a Z-test of proportions can be used. For situations when there are three or more levels of the discrete independent variable, the chi square test of independence should be used (Sect. 6.5.2).

5.8.1   One-Sample Z-Test of Proportions

With Z-test the outcomes have two discrete levels (e.g., pass/fail; yes/no; success/failure). Sample results are compared to the anticipated proportion for the population being tested. The null and alternative hypotheses are:

$$H_0 : \quad \hat{p} = P_0$$
$$H_1 : \quad \hat{p} \neq P_0$$

where $\hat{p}$ is the proportional results for the sample and P_0 is the anticipate proportion based on previous results or history with a product or assay. The term $1 - P_0$ is the complement proportion to P_0 or the probability of "not" the expected proportion for the population.

Assume there is an expected success rate (e.g., 96%), but a sample of 30 tablets has a success rate of only 93.3% (two failures or $\hat{p} = 0.933$). Is this outcome significantly different from the proportion that would be expected (0.96)? If the result is significantly different, then the null hypothesis is rejected. As with other inferential statistics, tests can be performed as either confidence intervals or the ratio method (Fig. 5.38). With the confidence interval approach, does the interval include the possible outcome of 0.96? If so the results would indicate a failure to reject the null hypothesis and two failures would be an expected result. However, if 0.96 fell outside the confidence interval, then the null hypothesis is rejected, and a failure rate of two is unacceptable.

A ratio method could also be used (Fig. 5.38). As discussed in Sect. 3.6, the critical Z-value for 95% confidence is 1.96 (2.58 for 99% confidence). With the ratio method, a Z-value is calculated. A decision rule could be written as follows for 95% confidence: reject the null if the calculated Z-value is greater than +1.96 or less than −1.96. Once again, Minitab simplifies the process by doing the calculation and providing a Z-value and corresponding p-value. If the p-value is less than 0.05, reject the null hypothesis.

Z-test can be two-sided or one-sided. In the previous example, a two-tailed test was performed. If in a one-tailed test, the critical value would be 1.64. A significant result of greater +1.64 or less than $a - 1.64$ would depend on the direction on the null hypothesis.

The one-sample Z-test is an excellent example of the importance of sample size. Assume the expected P_0 is 0.50 (e.g., a coin toss). Figure 5.39 shows the results for the same proportional differences with an increasing number of observations; in this case $\hat{p} = 0.65$ (13 out of 20 tosses are heads). Note that if these results appeared with more than 47 or 48 tosses, the results would be significant at 95% confidence, and the null hypothesis would be rejected. If the same proportional difference exists with over 75 tosses, H_0 can be rejected with 99% confidence that the coin is unfair ($P_0 \neq 0.50$).

One-sample Z-test of proportions (confidence interval assuming a normal distribution):

$$P_O = \hat{p} \pm Z_{(\alpha/2)} \sqrt{\frac{\hat{p}(1 - \hat{p})}{n}}$$

Two-sample Z-test of proportions (confidence interval):

$$P_1 - P_2 = (\hat{p}_1 - \hat{p}_2) \pm Z_{(\alpha/2)} \sqrt{\frac{\hat{p}_1(1 - \hat{p}_1)}{n_1} + \frac{\hat{p}_2(1 - \hat{p}_2)}{n_2}}$$

General equation for ratio tests:

$$Z = \frac{difference\ between\ proportions}{standard\ erro\ of\ the\ difference\ of\ the\ proportions}$$

One-sample Z-test of proportions (ratio method):

$$z = \frac{\hat{p} - P_0}{\sqrt{\frac{P_0(1 - P_0)}{n}}}$$

Two-sample Z-test of proportions (ratio method):

$$z = \frac{\hat{p}_1 - \hat{p}_2}{\sqrt{\frac{\hat{p}_1(1 - \hat{p}_1)}{n_1} + \frac{\hat{p}_2(1 - \hat{p}_2)}{n_2}}}$$

Fig. 5.38 Equations for the Z-tests of proportion

Fig. 5.39 Effects of sample size on Z-test results

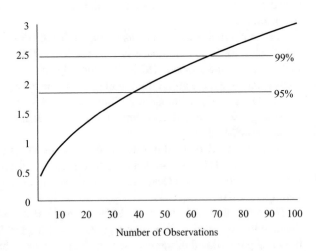

5.8.2   Two-Sample Z-Test of Proportions

Expanding the one-sample case, it is possible to compare two proportions, for example, proportions from two different assays (levels of an independent variable). In this case the hypotheses are:

$$H_0 : P_1 = P_2$$
$$H_1 : P_1 \neq P_2$$

The hypotheses make statements about the populations (P_1 and P_2), but the inferential statistic is based on the sample results ($\hat{p}_1$ and $\hat{p}_2$). Equations for both the one-sample and two-sample tests are presented in Fig. 5.38 and calculated as either a confidence interval or ratio method. The hypotheses also can be modified to perform a one-tailed or two-tailed test on the proportions. As in the previous test, the Z-statistic is based on the normalized standard distribution (Z-distribution in Table B1, Appendix B), with significant values being 1.96 for 95% confidence (Minitab default level), 1.64 for 90%, and 2.58 for 99% confidence. Like the one-sample test, Minitab helps by providing the Z-value and corresponding p-value. If the p-value is greater than 0.05, fail to reject the null hypothesis of equal proportions. Examples of both the one-sample and two-sample cases will be presented below.

5.8.3   Power and Sample Size Determination for Z-Tests of Proportion

For both the one-sample and two-sample Z-test of proportions, the appropriate sample size and amount of statistical power can be determined. Minitab can perform these calculations, and the statistics for these determinations are provided by Zar (1999) (pp. 558–561) and De Muth (2014) (pp. 397–399).

5.8.4   Minitab Applications

Steps for the one-sample and two-sample Z-test of proportions are can used for results expressed as proportions. One-sample test evaluates success/failure sample outcomes compared to an expected population. The two-sample case compares success/failure results between two levels of a discrete independent variable. Power and sample size determinations are available for both tests. Steps for performing these tests are presented below.

5.8.4.1   One-Sample Z-Test of Proportions

Procedure   Stats ➜ Basic Statistics ➜ 1 Proportion

Data input   There are two types of input: (1) column with dichotomous results
                 (e.g., y/n) or (2) summary results where the "Number of events"
                 (number of outcomes of interest) and "Number of trails" (total
                 sample size) are entered.

Hypothesis test On the dialog box if "Preform hypothesis test" is checked, an
                 anticipated population hypothesized proportion if required. If not
                 check the proportion defaults to 0.50.

Options   Automatic default for a 95% confidence interval for a two-tailed
                 test. These can be changed to a different degree of type I error if
                 required or a one-tailed test is desired. Also use the "Normal
                 approximation" method rather than the "Exact" method to
                 approximate formula in Fig. 5.38.

Report   Will produce both a confidence interval and p-value.

Interpretation   If the hypothesized proportion falls within the confidence interval
                 or the p-values is greater than 0.05, fail to reject the null
                 hypothesis.

5.8.4.2   Sample Size and Power Determination for a One-Sample Z-Test of Proportions

Procedure   Stat ➜ Power and Sample Size ➜ 1 Proportion

Data input   The estimated population proportion is required and two of the
                 following three: (1) sample size, (2) comparison proportion, and
                 (3) power value. Program will solve for the third item not entered.
                 Multiple values can be selected for any one of the two input items
                 (e.g., sample sizes of 10, 20, 30, and 40).

Options   Automatic default for a 95% confidence interval for a two-tailed
                 test. These can be changed to a different degree of type I error if
                 required or a one-tailed test is desired.

Graphs   By default will present a power curve, which can be turned off
                 under "Graph." Multiple sample sizes can be added to create a
                 graph with multiple curves and power points.

Report   Results for the third item not entered (e.g., if sample size and
                 detectable difference are entered, results will report the power).
                 Multiple entries (e.g., sample size) will report results for each
                 sample size entered.

Interpretation   Identifies the power for different sample sizes or different detect-
                 able differences or the required sample size for a desired power
                 and detectable difference.

5.8.4.3   Two-Sample Z-Test of Proportions

Procedure	Stats → Basic Statistics → 2 Proportion
Data input	There are three types of input: (1) data with each column a variable on the worksheet ("Samples" is the dependent variable, and "Sample IDs" is the independent variable); (2) levels of the independent variable on different columns in the worksheet; or (3) summary results where the "Number of events" (outcomes of interest) and "Number of trails" (total sample size) are entered.
Options	Automatic default for a 95% confidence interval for a two-tailed test. These can be changed to a different degree of type I error if required or a one-tailed test is desired. There are two "Methods" available: (1) estimating the proportions separately or (2) using a pooled estimate of the population. The formula in Fig. 5.38 represents separate proportions. Results may differ slightly depending on which method is chosen.
Report	Will provide both a confidence interval and p-value.
Interpretation	If zero falls within the confidence interval or the p-values is greater than 0.05, fail to reject the null hypothesis. Use the p-value associated with Z-value unless there is very small sample size, in which case the Fisher p-value would be used (see Chap. 6).

5.8.4.4   Sample Size and Power Determination for a Two-Sample Z-Test of Proportions

Procedure	Stat → Power and Sample Size → 2 Proportion
Data input	The results for one of the proportions are required ($p2$) and two of the following three: (1) sample size, (2) the second proportion ($p1$), and (3) power. Program will solve for the third item not entered. Multiple values can be selected for any one of the two input items (e.g., sample sizes of 10, 20, 30, and 40).
Options	Automatic default for a two-tailed test and a type I error (α) of 0.05 can select an alternative one-tailed test or different α (significance level).
Graphs	By default will present a power curve, which can be turned off under "Graph." Multiple sample sizes can be added to create a graph with multiple curves and power points.
Report	Results for the third item not entered (e.g., if sample size and detectable difference are entered, results will report the power). Multiple entries (e.g., sample size) will report results for each sample size entered.

Interpretation   Identifies the power for different sample sizes or different detectable differences or the required sample size for a desired power and detectable difference.

Interpretation   The easiest interpretation is to look at the reported p-value. If the p is less than 0.05, reject the null hypothesis in favor of the alternative hypothesis; if not, fail to reject the null hypothesis.

5.8.5   Examples

Historically during production runs for a specific oral tablet, there is an expected defect rate (minor blemish or chipping) of 1.2%. During one specific run, a sample of 100 tablets were found to have a defect rate of 5% (5 defective tablets). Does this result differ significantly from what would normally be expected for this product? Here the P_0 is expected to be 0.012, and the sample result ($\hat{p}$) is 0.05 based on an $n = 100$ observations. This case would appear to be appropriate for a one-sample Z-test of proportions and with Minitab that it is easier to use the "Summarized data" option on the dialog box, where the trials are 100, the number of events (defects) is 5, and hypothesized proportion is 0.012. The results of the analysis are presented in Fig. 5.40. In this case the sample creates a 95% confidence interval of 0.007 to 0.093 and would include the expected amount of defects (0.012). However the ratio method produces a $p = 0.007$, and the researcher would reject the null hypothesis and conclude that the batch from where the sample was taken had a higher defect rate than historically expected. A control table could be created to give the operator

Fig. 5.40 Minitab output for the example of a one-sample Z-test of proportions

Confidence interval (with normal approximation):

N	Event	Sample p	95% CI for p
100	5	0.050000	(0.007284, 0.092716)

Ratio method:

Null hypothesis	H_0: p = 0.012
Alternative hypothesis	H_1: p ≠ 0.012

Z-Value	P-Value
3.49	0.000

p-values associated with different numbers of defects:

Number of defects	p-value
0	0.270
1	0.854
2	0.463
3	0.098
4	0.010
5	<0.001
6	<0.001

an indication of acceptable defect rates. As seen in the bottom of Fig. 5.40 anytime there are four or more defects, the results are significantly greater than the expected number of defects is based on a sample size of 100.

The researcher wanted to be able to detect at least a 20% variation from the expected proportion of defects if such a difference existed. How large a sample would be required? In this case a 20% variation would be 0.0024 (0.20×0.012). Using Minitab the power of detecting a 20% difference was only 0.008 (Fig. 5.41). The required number of samples to detect this large difference with 80% power would be 704 samples from a batch.

In a second example, samples are taken during the development of a final dosage form, and the frequency of defects was analyzed to determine the effect of the speed of the tablet press. Samples were collected at 80,000 (lower) and 120,000 (higher) units per hour. Initially 500 tablets were to be collected at each speed; unfortunately due to an accident, only 480 tablets were retrieved at the higher speed. The number of defects at the higher speed was 17 and only 11 at the lower speed. The researcher would like to know if there is a significant difference between defect rates at the two tablet press speeds. Once again the easiest approach is to use the "Summarized data" option on the dialog box, where the trials for the slower speed are 500 with a corresponding event of 12 and 480 trials for the faster speed with 11 events. The results of the analysis are presented in Fig. 5.42. For this example the null hypothesis that the two speeds produce the same proportion of defects cannot be rejected. For the 95% confidence interval, zero falls outside the interval (-0.0344 to $+0.0075$) and the $Z = -1.26$, $p = 0.209$, which is greater than 0.05. The researcher can state that she was unable to identify a significance difference in the proportion of defects based on the speed of the tablet press. To reject the null hypothesis, the researcher would need to accept 20.9% chance of being wrong with that decision.

Was the sample size adequate? Of concern to the researcher is the ability to detect a 5% difference if it exists at the two speeds. Using the lower probability 0.022 (11/500), what is the power to detect a difference if the second proportion is 0.072? In this example the power to detect such a difference was 0.956 (Fig. 5.43) well in excess of the goal of 0.80. The required number of sample to reach a power of 0.80 would only require 281 samples for each tablet press speed.

Power given 100 observations to detect a 3% difference :

Comparison p	Sample Size	Power
0.0024	100	0.0082081

Required observations for power of 0.80 to detect a 3% difference:

Comparison p	Sample Size	Target Power	Actual Power
0.0024	704	0.8	0.800710

Fig. 5.41 Minitab output for Sample size and power determinations for the one-sample Z-test of proportions example

Confidence interval:

	95% CI for
Difference	**Difference**
0.0134167	(-0.007529, 0.034362)

CI based on normal approximation

Ratio method:

Null hypothesis	H_0: $p_1 - p_2 = 0$	
Alternative hypothesis	H_1: $p_1 - p_2 \neq 0$	
Method	**Z-Value**	**P-Value**
Normal approximation	1.26	0.209
Fisher's exact		0.251

Sample	**N**	**Event**	**Sample p**
Sample 1	480	17	0.035417
Sample 2	500	11	0.022000

Fig. 5.42 Minitab output for the example of a two-sample Z-test of proportions

Power given 480 observations to detect a 5% difference :

	Sample	
Comparison p	**Size**	**Power**
0.072	480	0.956553

The sample size is for each group.

Required observations for power of 0.80 to detect a 3% difference:

	Sample	**Target**	
Comparison p	**Size**	**Power**	**Actual Power**
0.072	281	0.8	0.801312

The sample size is for each group.

Fig. 5.43 Minitab output for sample size and power determinations from the two-sample Z-test of proportions example

References

Allen DM (1974) The relationship between variable selection and data augmentation and a method for prediction. Technometrics 16(1):125–127

Box GE, Hunter WG, Hunter JS (1978) Statistics for experimenters. John Wiley and Sons, New York

Conover WJ (1999) Practical nonparametric statistics. John Wiley and Sons, Inc., New York

Daniel WW (1978) Applied nonparametric statistics. Houghton Mifflin, Boston

De Muth JE (2014) Basic statistics and pharmaceutical statistical applications, 3rd edn. CRC Press, Boca Raton, FL

Dunnett CW (1955) A multiple comparison procedure for comparing several treatments with a control. J Am Stat Assoc 50(2):1096–1121

Hollander M, Wolfe DA (1999) Nonparametric statistical methods, 2nd edn. John Wiley and Sons, New York

Hsu JC (1992) Stepwise multiple comparisons with the best. J Statist Plann Inference 33(2):197–204

Kruskal WH, Wallis WA (1952) Use of ranks on one-criterion variance analysis. J Am Stat Assoc 47(620):583–612

Mann HB, Whitney DR (1947) On a test of whether one of two random variables is stochastically larger than the other. Ann Math Stat 18(1):50–60

Mood AM (1950) Introduction to the theory of statistics. McGraw-Hill, New York

Pearson ES, Hartley HO (1970) Biometrika tables for statisticians, Vol. 1 (Table 29). Biometrika Trustees at the University Press, Cambridge/London

Petersen RG (1985) Design and analysis of experiments. Marcel Dekker, New York

Salsburg D (2002) The lady tasting tea: how statistics revolutionized science in the twentieth century. Henry Holt and Company, New York

Satterthwaite FE (1946) An approximate distribution of estimates of variance components. Biom Bull 2(6):110–114

Student (1908) The probable error of a mean. Biometrika 6(1):1–25

Zar JH (1999) Biostatistical analysis, 4th edn. Prentice Hall, Upper Saddle River

Chapter 6
Multivariate Analysis: Tests to Evaluate Relationships

Abstract Sometimes a researcher wishes to determine if variables are associated with each other. If one variable increases, what is the effect on a second variable? This chapter explores inferential statistical tests that evaluate the relationship between two or more variables. The tests will determine the strength of the relationship and if the relationship is strong enough to show statistical significance. Selection of the test will depend on whether the variables are continuous or discrete. For continuous variables, tests covered in this chapter include the correlation, linear regression, multiple regression, and logistic regression. For discrete variables, chi square and related tests are discussed including tests for goodness-of-fit and independence. Additional tests related to chi square are introduced in the chapter. Minitab applications for evaluating the strength of these relationships are presented.

Keywords Coefficient of determination · Chi Square · Confidence bands · Correlation · Goodness-of-fit · Independence · Linear regression · Logistic regression · Multiple regression · Prediction bands

In contrast to the previous chapter on inferential tests used to identify differences, this chapter explores tests for relationships between variables. With correlation the researcher is concerned with the relationship among two or more continuous dependent variables. However, with regression analysis there is at least one independent continuous variable that the researcher can control in the study. Also discussed will be relationships where the outcomes are discrete; where the chi square and associated tests can be used analyze results. These may involve an independent variable or could evaluate the relationship among multiple discrete dependent variables. In many cases these could be considered predictive statistics where the independent variable can be used to make a prediction for the dependent (response) variable.

© American Association of Pharmaceutical Scientists 2019
J. E. De Muth, *Practical Statistics for Pharmaceutical Analysis*, AAPS Advances in the Pharmaceutical Sciences Series 40,
https://doi.org/10.1007/978-3-030-33989-0_6

6.1 Correlation

Correlation does not require an independent (predictor) variable. With correlation, two or more dependent variables may be compared to determine if there is a relationship and to measure the strength of that relationship. Correlation describes the degree to which two or more sample variables show interrelationships representing their given population. The amount of relationship is expressed by the *correlation coefficient* (*r*), and the equation is presented in Fig. 6.1. This *r*-value, sometime referred to as the *Pearson correlation coefficient* or *product-moment correlation coefficient*, may be either positive or negative, and values can vary for a perfect negative correlation of −1.0 to a perfect positive correlation of +1.0.

The null hypothesis would be that no correlation exists:

$$H_0 : \quad r_{yx} = 0$$

$$H_1 : \quad r_{yx} \neq 0$$

where zero represents no relationship between two dependent variables. The alternative hypothesis is that zero cannot exist and a significant relationship existed between the two variables. Results are based in part on the sample size, and critical values for correlation coefficients are presented in Table B6 (Appendix B). Most software packages will list both the *r*-value and a corresponding *p*-value. As with previous test, *p*-values less than 0.05 are usually considered statistically significant. Such a significant correlation does not explain why the relation occurs, only that such a relationship exists. For example, in a clinical trial it is found that renal clearance is correlated with blood pressure; clearance is positively correlated with blood pressure. Does lower renal clearance cause lower blood pressure or does blood pressure affect renal clearance or could there be a third factor? A significant correlation coefficient cannot answer this question.

Pearson Correlation:

$$r = \frac{\Sigma(x - \bar{X}_x)(y - \bar{X}_y)}{\sqrt{\Sigma(x - (\bar{X}_x)^2(y - (\bar{X}_y)^2}}$$

Equivalent computational formula:

$$r = \frac{n\Sigma xy - \Sigma x \Sigma y}{\sqrt{n\Sigma x^2 - (\Sigma x)^2}\ \sqrt{n\Sigma y^2 - (\Sigma y)^2}}$$

where *xy* is the product of *x* and *y* for each pair

Spearman rho:

$$\rho = 1 - \frac{6(\Sigma d^2)}{n^3 - n}$$

Where *d* is the difference between the ranks

Fig. 6.1 Equations for a correlation coefficient and nonparametric procedure

A significant positive or negative correlation between two variables may show that a relationship exists. Whether one considers it as a strong or weak correlation, important or unimportant, is a matter of interpretation. For example in a clinical trial a correlation of +0.80 would be considered a high correlation. However, individuals in the pharmaceutical industry doing a process validation may require a correlation at least +0.999.

Verbal descriptions of correlations are inconsistent. The simplest might be less than 0.25 is a "doubtful" correlation; 0.26–0.50 represents a "fair" correlation; 0.51–0.75 is a "good" correlation; and greater than 0.75 can be considered a "superior" correlation (Kelly et al. 1992). Another rough guide is less than 0.20 is slight, almost negligible relationship; 0.20–0.39 is a low correlation, definite but small relationship; 0.40–0.69 represents a moderate correlation, substantial relationship; 0.70–0.90 is a high correlation with a marked relationship; and greater than 0.90 is considered a very high correlation and a very dependable relationship (Guilford 1956). The sign (+ or −) would indicate a positive or negative correlation. For example for $r = +0.954$ would be considered a very high correlation with a very dependable relationship between the two variable populations based on the sample data.

One graphic way to evaluate a relationship between two variables is a scatter plot which presents data points on an x- and y-axis (Fig. 6.2). Unlike linear regression (discussed below), correlations do not define a line, but indicate how close the data is to falling on a straight line. If all the data points were aligned in a straight diagonal, the correlation coefficient would equal to a +1.0 or −1.0. As a correlation approaches +1.0 or −1.0, there is a tendency for numbers to concentrate closer to a

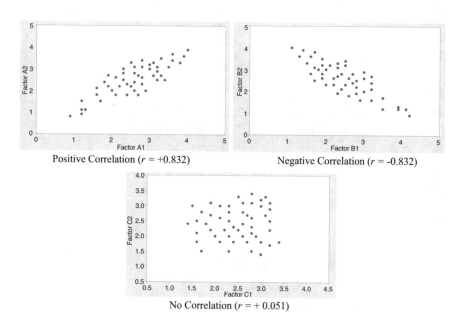

Positive Correlation ($r = +0.832$)   Negative Correlation ($r = -0.832$)

No Correlation ($r = + 0.051$)

Fig. 6.2 Examples of graphic presentations of correlation data

straight line. However, one should not assume that just because there is a "high" or "strong" correlation that it necessarily follows a straight line. Figure 6.3 illustrates this point in a classic set of data by Anscombe (1973). In this figure four different data sets all produce the same "high" correlation ($r = 0.816$); however, their distributions are completely different. As discussed below with linear regression, if a line were drawn that best fit between the points in each data set, they would be identical with a slope of 0.5 and a y-intercept of 3.0. This figure also shows the advantage of plotting the data on graph paper, or creating a computer-generated visual, to actually observe the distribution of the data points.

The correlation coefficient says nothing about the percentage of the relationship, only its relative strength. For example a correlation of 0.90 is not twice as strong a relationship as one for a correlation of 0.45. It represents a convenient ratio, not an actual measurement scale. It serves primarily as a data reduction technique and as a descriptive method. As with other statistical ratios (e.g., two-sample t-tests), what is important (differences between observations and mean on both axis) is the numerator and a standard error term is in the denominator.

One needs to be cautious of correlation and *causality* (the relationship between cause and effect). The correlation coefficient does not suggest nor prove the reason for this relationship; only that it exists. It determines whether the two variables differ together either positively or negatively and the degree of this relationship. It does not indicate anything about what causes this relationship since neither variable is

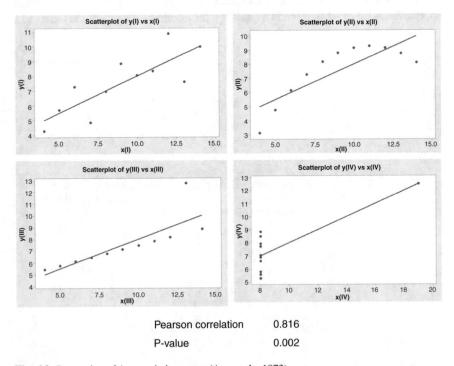

| Pearson correlation | 0.816 |
| P-value | 0.002 |

Fig. 6.3 Recreation of Anscombe's quartet (Anscombe 1973)

controlled by the researcher. Did the x-variable cause the result in y? Did y affect variable x? Could a third variable have affected both x and y? There could be many reasons for a significant relationship.

Correlation is a parametric test and as such assumes that the population represented by both the x- and y-axes are normally distributed. If one or both of the axis represent population that are not normally distributed, then either a transformation of the data is required or an alternative nonparametric procedure such as a Spearman rho should be used. With correlation the variables may be measured on different scales (e.g., weight vs. height); therefore, homogeneity of variance is not necessarily a requirement.

6.1.1 Multiple Correlations

Two dependent variables are used to create a single correlation coefficient, but correlations can be calculated for more than just two dependent variables. In this case software usually creates a matrix of all pairwise relationships (comparison only two of the possible variables) and whether or not they are significant. This matrix will usually reported with a series of r-values for each comparison and corresponding p-values (see example below). Once again the easiest way to determine significant relationships is to look for p-values less than 0.05 and equally important whether the sign is positive or negative indicating the direction of that relationship.

An alternative method for the evaluation of multiple correlations is to calculate a *partial correlation coefficient* that shows the correlation between two continuous variables, while removing the effects of any other continuous variables that may confound the results. This test is beyond the scope of this book and not available with Minitab. Additional information about the partial correlation coefficient is available (De Muth 2014, pp. 330–331).

6.1.2 Nonparametric Alternative

The most common nonparametric alternative to the Pearson correlation coefficient is the *Spearman rank-order correlation*. Also referred to as the *Spearman rho*, the null hypothesis is the same as correlation, namely, that no relationship exits:

$$H_0: \quad \rho = 0$$
$$H_1: \quad \rho \neq 0$$

The calculation involves the differences between the rankings for the two dependent variables (Fig. 6.1). Determination of significance is based on the p-value. This test is especially useful for looking at relationships between ordinal data (Spearman 1904).

6.1.3 Minitab Applications

With Minitab it is possible to do a simple correlation with two dependent variables, more complex matrices with more than two dependent variables, visualize the relationship with scatter plots or use a nonparametric Spearman rho statistic to evaluate the strength of relationships. Steps for performing these tests are presented below.

6.1.3.1 Correlation Coefficient

Procedure	Stats → Basic Statistics → Correlation
Data input	Select columns from the worksheet.
Options	Select "Pearson correlation" in the top method box.
Graphic	Will automatically produce a scatter plot unless the option is turned off. For "statistics to display on plot" options include correlation and the p-value.
Results	It is important to check the pairwise correlation table in order to obtain the associated p-value.
Report	Both the correlation coefficient and a p-value are displayed (unless pairwise correlation table is turned off).
Interpretation	Determine the direction of the correlation based on the sign (positive values do not display + sign). The importance of the correlation coefficient (r) is determined by the p-value with anything less than 0.05 being a significant relationship.

6.1.3.2 Scatter Plot

Procedure	Graph → Scatter Plot
Initial decision	Select the type of plot required (simple, with groups, with lines).
Data input	Select columns from the worksheet for the y-axis and x-axis.
Options	Multiple options are available for design and labeling of the scatter plot.
Report	Create a scatter plot for the two columns selected.

6.1.3.3 Nonparametric Alternative: Spearman Rho

Procedure	Stats → Basic Statistics → Correlation
Data input	Select columns from the worksheet.
Options	Select "Spearman correlation" in the top method box.
Graphic	Will automatically produce a scatter plot unless the option is turned off. For "statistics to display on plot" options include correlation and the p-value.

Results   It is important to check the pairwise correlation table in order to
                   obtain the associated p-value.
Report   Both the correlation coefficient and a p-value are displayed
                   (unless pairwise correlation table is turned off).
Interpretation   Determine the direction of the correlation based on the sign (pos-
                   itive values do not display + sign). The importance of the
                   Spearman rho (ρ) is determined by the p-value with anything less
                   than 0.05 being a significant relationship.

6.1.4   Examples

Two scales are used to measure certain analytical outcome. Method A is an established test instrument, while Method B (which has been developed by the researchers) is quicker and easier to complete. The results of the initial study are presented
in Table 6.1. Is there a correlation between the two measures? The evaluation of the
sample data is presented in Fig. 6.4. There is a very strong positive correlation
between the two methods ($r = +0.992, p < 0.001$). Depending on regulatory or internal standards, this might be acceptable for method transfer. Graphically the data
appears to fall close to a straight line. Similar results are seen with the Spearman rho
correlation ($\rho = 0.996, p < 0.001$).

   As a second example, a study was conducted to compare the percent of hepatitis
C virus neutralized to E2 measures based on three different test kits. The results of
the study are presented in Table 6.2. Based on these results of the study, which test
offers the highest correlation between the E2 measures and neutralized HCV? The
resulting matrix in Fig. 6.5 represents all pairwise comparisons. Of major interest to
the investigators are the results in the first column on the left comparing the HCV
neutralizing with each of the three kits being tested. The upper number in each pairing represents the r-value and the lower one the p-value associated with that correlation coefficient. It appears the test kit C provides the strongest correlation
($r = +0.791, p < 0.001$) with a very high correlation and a very dependable relationship. If the researcher was concerned about normality in the population for either
the x- or y-axis, a nonparametric alternative could be performed. In this case
Spearman rho tests provide similar results (Fig. 6.5) with test kit C once again providing the best correlation ($\rho = 0.690, p < 0.001$). Graphically (Fig. 6.6) it can be
seen why test kit C has the strongest correlation.

6.2   Linear Regression

Unlike the correlation coefficient, regression analysis requires at least one independent researcher controlled variable. Where correlation describes pairwise relationships between continuous variables, linear regression (also termed *regression*

Table 6.1 Results comparing
two analytical methods

Sample	Method A	Method B
1	55	56
2	66	67
3	46	45
4	77	75
5	57	57
6	59	59
7	70	69
8	57	59
9	52	51
10	36	38
11	44	45
12	55	56
13	53	51
14	67	68
15	72	74

Correlation coefficient:

Sample 1	Sample 2	Correlation	95% CI for ρ	P-Value
Method B	Method A	0.992	(0.976, 0.998)	0.000

Scatter plot:

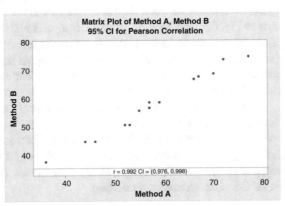

Spearman rho:

Sample 1	Sample 2	Correlation	95% CI for ρ	P-Value
Method B	Method A	0.996	(0.982, 0.999)	0.000

Fig. 6.4 Minitab output and graphic for the correlation between two analytical methods

analysis to cover a variety of tests) is a statistical method to evaluate how one or more independent (predictor) variables influence outcomes for one continuous dependent (response) variable. Both linear regression and correlation describe the strength of the relationship between two or more continuous variables. However, with linear regression, a relationship is established between the two variables and a response for the dependent variable can be estimated based on a given value for the independent variable. For correlation, two dependent variables can be compared to determine if a relationship exists between them. In regression analysis researchers control the values of at least one of the variables and evaluated outcomes at different levels of these variables. Where correlation simply describes the strength and direction of the relationship, regression analysis provides a method for describing the nature of the relationship between two or more continuous variables.

In linear regression a line is computed that best fits between the data points. If a linear relationship is established, the magnitude of the effect of the independent variable can be used to predict the corresponding magnitude of that effect on the dependent variable. For example a particulate diameter can be used to predict surface area. The strength of the relationship between the two variables can be determined by calculating the amount of the total variability that can be accounted for by the regression line.

Table 6.2 Results comparing three test kits for E2 measures and percent of neutralized HCV

% HCV neutralized	E2 measurements		
	Test kit A	Test kit B	Test kit C
94.62	0.776	0.273	1.673
94.27	0.587	0.060	1.663
85.45	0.620	0.268	1.572
99.09	2.098	0.196	2.261
94.15	0.583	0.270	1.641
67.07	0.341	0.029	1.207
97.85	1.283	0.701	1.933
56.18	0.261	0.001	0.781
64.66	0.450	0.026	1.063
96.75	1.036	0.184	1.913
96.47	0.928	0.467	1.900
84.63	0.572	0.032	1.560
94.16	0.583	0.270	1.641
96.04	1.278	0.134	1.886
60.07	0.418	0.025	1.027
95.55	1.059	0.282	1.824
83.84	0.548	0.260	1.553
95.39	0.804	0.129	1.790
81.44	0.521	0.029	1.496
77.67	0.349	0.237	1.253

Correlation coeffient:

	%HCV Neut.	Test A	Test B
Test A	0.444		
	0.000		
Test B	0.082	0.079	
	0.532	0.548	
Test C	0.791	0.325	0.144
	0.000	0.011	0.271

Cell Contents
Pearson correlation
P-Value

Spearman rho:

	%HCV Neut.	Test A	Test B
Test A	0.543		
	0.000		
Test B	0.047	-0.005	
	0.719	0.969	
Test C	0.690	0.254	0.068
	0.000	0.050	0.606

Cell Contents
Spearman rho
P-Value

Fig. 6.5 Minitab output for Pearson correlations and Spearman rho tests comparing three test kits with HCV neutralization

There are several assumptions associated with the linear (two-dimensional) regression. First, values on the x-axis, which represent the independent variable are "fixed," meaning points are chosen by the researcher (e.g., in a stability study samples are taken a 6, 12, 18, 24, 36, and 48 months, not a random time points). These nonrandom predetermined points in the study are compared to responses on the y-axis. Because the researcher controls the x-axis, it is assumed that these measures are without error. Second, for each value on the x-axis, there is a subpopulation of values (represented by sample data) for the corresponding dependent variable on the y-axis. It is assumed that these subpopulations are normally distributed in the population (linear regression is a parametric test). For data that may not be normally distributed, for example, AUC or Cmax measures in bioavailability studies, log transformations may be required to convert such positively skewed data to a more normally distributed subpopulation. Coupled with the assumption of normality is homogeneity of variance. It is assumed that the variances for all

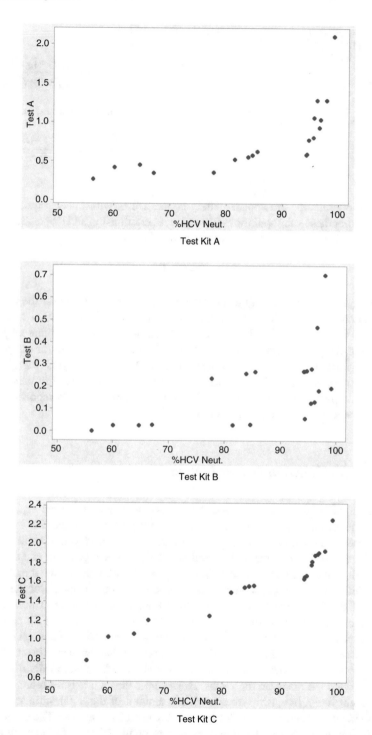

Fig. 6.6 Minitab graphics of the relationship between different test kits and HCV neutralization

Fig. 6.7 Equations for   Regression line: $y = a + bx$
regression line slope (b)
and intercept (a)

$$b = \frac{n \sum xy - (\sum x)(\sum x)}{n \sum x^2 - (\sum x)^2}$$

where xy is the product of x and y for each pair

$$a = \frac{\sum y - b \sum x}{n}$$

these subpopulations are approximately equal. Third, with linear regression it is assumed that these subpopulations have a linear relationship and that a straight line can be drawn between them. The formula for this line is

$$\mu_{y/x} = \alpha + \beta x$$

where $\mu_{y/x}$ is the mean for any given subpopulation for an x-value for the predictor independent variable. The terms α and β represent the true population y-intercept and slope for the regression line, respectively. Unfortunately, the population parameters are usually unknown, but the *regression line* can be estimated for sample data using the formula $y = a + bx$. This regression line has the synonyms *line of best fit* or the *least squares line* (the line with the least amount of variability between the lines a data points on a vertical axis). Computer programs usually report the slope and intercept using the equations in Fig. 6.7.

6.2.1   Coefficient of Determination

As the spread of the scatter dots along the vertical axis (y-axis) decreases, the precision of the estimated μ_y increases. A perfect (100%) estimate is possible only when all the dots (data points) lie on the straight regression line. The coefficient of determination (r^2) offers one method to evaluate if the linear regression equation adequately describes a linear relationship or a linear *model*. It compares the scatter of data points about the regression line with the scatter about the mean for the sample values of the dependent y-variable. Figure 6.8 shows a scattering of points about both the mean of the y-distribution ($\bar{X}_y$) and the regression line itself for part of the data presented in the figure. It is possible to measure the deviation of each point (y_i) from the mean on the y-axis (labeled C in Fig. 6.8). If there were no linear relationship between the x- and y-variables, one would expect a random distribution of points around the mean on the y-axis. However, if the data is truly represented by the straight line relationship, then a certain amount of this total variation can be explained by the deviation from the mean to the line (B in Fig. 6.8). The point on the straight line is labeled y_c. However, most data points will not fall exactly on the

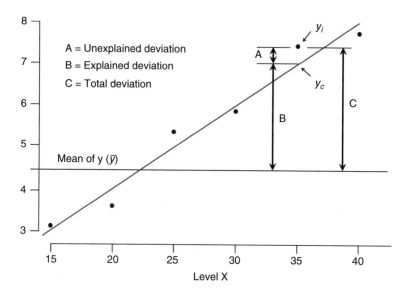

Fig. 6.8 Variability of data points around the mean for the y-variable and the regression line

regression line, and this deviation (*A* in Fig. 6.8) must be caused by other source(s) (random error). The coefficient of determination is calculated using the sum of the squared deviations that takes into consideration these deviations (*A*, *B*, and *C*). In this case the total deviation equals the explained deviations (defined by the line) plus the unexplained deviations, where the total deviation is the vertical difference between the observed data points and the mean for the y-axis ($y_i - \bar{y}$). The explained deviation is the vertical difference between the points on the regression line and the mean for the y-axis ($y_c - \bar{y}$). The unexplained deviation is the vertical difference between the observed data points and their corresponding points on the regression line ($y_c - y_i$). These vertical distances between the data points and the regression line are called *residuals* (*A* in Fig. 6.8). In a perfect situation the regression line between the data points, the sum of the residuals should equal zero, an equal amount of deviation above and below the line. Thus, the regression line is the line that results in the smallest value for the sum of the squared deviations, $\sum(y_c - y_i)^2$. This term is referred to as the *residual sum of squares* or *error sum of squares*. This will become the error term in an ANOVA table that will be used to determine the significance the regression analysis. Calculations for these sums of squares associated with the explained, unexplained, and total variability in a regression analysis are shown in Fig. 6.9.

The coefficient of determination (r^2) is the proportion of variability accounted for by the sum of squares due to linear regression. It is the ratio of the variability that is accounted for by the regression line divided by the total amount of variability (Fig. 6.9). The coefficient of determination measures the amount of fit of the regression equation to the observed values for *y*. In other words, the coefficient of determination identifies how much variation in the dependent variable can be

Sum of Squares for ANOVA table

1. Sum of squares total (SS_T)

$$SS_T = \sum (y_i - \bar{y})^2 = \sum y^2 - \frac{(\sum y)^2}{n}$$

2. Sum of squares explained (SS_E)

$$SS_E = \sum (y_c - \bar{y})^2 = b^2 \times \left[\sum x^2 - \frac{(\sum x)^2}{n} \right]$$

3. Sum of squares unexplained (SS_U)

$$SS_U = \sum (y_c - y_i)^2 = SS_T - SS_E$$

Coefficient of determination (r^2)

$$r^2 = \frac{SS_E}{SS_T} = \frac{b^2 \times \left[\sum x^2 - \frac{(\sum x)^2}{n} \right]}{\sum y^2 - \frac{(\sum y)^2}{n}}$$

Adjusted coefficient of determination ($R^2(adj)$)

$$R^2_{adj} = 1 - \left[\frac{(1 - R^2)(n - 1)}{n - k - 1} \right]$$

Fig. 6.9 Equations used to calculate the coefficient of determination

explained by variations in the independent variable. Multiplying the r^2 by 100, it is possible to express the percent of total variability on the y-axis (dependent variable) accounted for by the regression line. For example, if the r^2 is 0.987, then the regression line accounts for 98.7% of all the variability in the dependent variable. Minitab expresses the results for the coefficient of determination as R^2 and as the percent.

The rest of the variability $(1 - r^2)$ is explained by other factors, most likely unidentifiable, random error unknown to the researcher, called the *coefficient of non-determination*.

Sometimes termed the *common variance*, r^2 represents that proportion of variance in the response (dependent) variable that is accounted for by variance in the predictor (independent) variable. As the coefficient of determination approaches 1.0, one is able to account for more of the variation in the dependent variable with values predicted from the regression equation. Obviously, the amount of error associated with the prediction of the response variable from the predictor variable will decrease as the degree of correlation between the two variables increases.

Therefore, the r^2 is a useful measure when predicting values for the dependent variable based on the independent variable.

Minitab provides additional information besides just the coefficient of determination. Printouts will also include S and $R^2(adj)$. S is the *standard error of the regression* or *standard error of the estimate* and approximates the accuracy of any prediction made by the regression line. The closer the S is to 1.00 the better the predictive model (in this case linear regression). The regression line minimizes the unexplained sum of squared deviations on the y-axis. This S is calculated as the square root of the mean squared residual off the ANOVA table. The $R^2(adj)$ is an adjusted coefficient of determination is defined in Fig. 6.9 and is considered unbiased estimate of the population coefficient of determination (R^2). The critical information for the determination of linearity is the ANOVA table with its F- and p-values presented in the next section.

It is important to understand that the linear regression line fits and defined the results between the data points and cannot be extrapolated beyond the largest or smallest point observed. As illustrated in Fig. 6.10, what happens after the largest recorded point on the x-axis? Would the linear relationship continue (A), might there be an acceleration in the effect (B), a leveling of response (C), or an actual decrease in the effect with increased values (D)? Correspondingly, it is not known what the relationship is for responses at dosages less than the smallest studied data point on the x-axis. If more data were available beyond the last data point, it might be found that the regression line would level out or decrease sharply. Therefore, the regression line and the regression equation apply only within the range of the x-values actually observed in the sample data. In Fig. 6.10 nothing can be predicted beyond 230 mg or below 175 mg which were the starting and ending points in the study design.

Fig. 6.10 Example of the problems associated with extrapolation

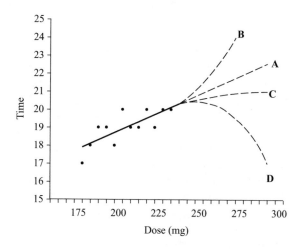

6.2.2   ANOVA Table for Linear Regression

Calculating the slope of the line it is possible to determine if there a strong positive or negative relationship between the two continuous variables and one can establish the type of relationship (linear or curvilinear). This final decision on the acceptability of the linear regression model is based on an objective ANOVA test where a statistical test will determine from sample data whether there is a significant linear relationship:

$$H_0 : \quad X \text{ and } Y \text{ are not linearly related}$$
$$H_1 : \quad X \text{ and } Y \text{ are linearly related}$$

This involves creating a new ANOVA table (Fig. 6.11) where the residuals become the error component. Similar to the analysis of variance table presented for a one-way ANOVA in Sect. 5.5, the important determinants are the F-value and its associated p-value. In addition to the table, Minitab provides supplemental information for the coefficient of determination, the adjusted R^2, and standard error of the regression.

6.2.3   Confidence and Prediction Bands

Without going through arduous mathematical equations, the following is a simple explanation on confidence bands. The computer can take sample data and calculate the regression line that fits best between the sample data points and defined the slope as b. This b is a sample outcome and the best guess of the true population slope β. Using an inferential statistic a confidence interval can be estimated for any point on the slope of the line based on sample data (similar to a one-sample t-test). So the line can pivot slightly around the mean on the x-axis, and the difference in possible regression lines would get wider at the extremes of the x-distribution. In this case the point is the best estimate for the confidence interval. A reliability coefficient is identified for a t-statistics, and a large error term is created based on how far the point is from the mean for values on the x-axis. Minitab and other statistical software packages can create these confidence intervals at any point on the continuum of the x-axis. These intervals are graphic presented as confidence bands (Fig. 6.12). Because the slope of the regression line can change, the width of the interval gets larger at the extremes and tightest at the mean on the x-axis. Confidence bands represent confidence interval of where population mean can be located (e.g., with 95% confidence the population mean is located somewhere between to estimate anywhere alone the regression line).

Prediction bands are similar to confidence bands but estimate an area of response for a single observation versus the population. This would be analogous to a confidence interval compared to tolerance limits (Sect. 3.7) for a single point on the x-axis. For the same sample results, the prediction bands will be wider than those for the confidence bands.

Source	Degrees of Freedom	Sum of Squares[1]	Mean Square	F	p
Linear Regression	1	SS_E	$\dfrac{SS_E}{1}$	$\dfrac{MS_{LR}}{MS_E}$	p-value[2]
Error (residual)	$n-2$	SS_U	$\dfrac{SS_U}{n-2}$		
Total	$n-1$	SS_T			

Example of a Minitab report:

Source	DF	SS	MS	F	P
Regression	1	5.8811	5.88112	14.28	0.004
Error	10	4.1189	0.41189		
Total	11	10.0000			

[1] Sum of squares are intermediates derived from computational formulas (Figure 6.9)

[2] p-value is the amount of type I error based on the F-statistic

Fig. 6.11 ANOVA table associated with linear regression

6.2.4   Nonlinear Situations

Regression analysis to this point has assessed the relationship between data and a straight line that fits between these points. Such information can be assessed using an ANOVA table or by the strength of the coefficient of determination (r^2). Visually, other graphic assessments of linearity include the plotting of the residuals. If a linear relationship exists, one would expect the residuals to be randomly distributed between positive and negative values. Residual plots (which are an option with Minitab) can be helpful in identifying possible nonlinear relationships. Figure 6.13 shows a residual plot for the illustrative example shown in Figs. 6.11 and 6.12. Notice the random nature of positive and negative residuals ($y_c - y_i$), this supports the assumption that a linear relationship exists.

Sometimes in research there are curvilinear relationships instead of simple linear relationships. Fortunately, many such relationships can also be expressed and evaluated as linear relationships. The goal of a curvilinear regression is to describe the shape of the relationship between two continuous variables. Is the best fit linear or could it be something else? Curvilinear regression is sometime called *trend analysis*. For curvilinear regression we use models to fit a curve instead of a straight line through our sample data points. Also referred to as *polynomial regression*, various polynomial equations are used to fit the curve. The most common are quadratic or cubic models,

Calculation for points on a confidence band:

$$\bar{y} = y_c \pm t_{n-2,1-\alpha/2} \times \sqrt{MS_{residual}} \sqrt{\frac{1}{n} + \frac{(x_i - \bar{X})^2}{\sum x^2 - \frac{(\sum x)^2}{n}}}$$

Example of confidence bands:

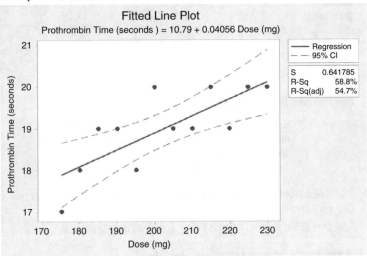

Fig. 6.12 Formula and graphic example of confidence bands for linear regression

which can be quickly accessed in Minitab. At the bottom of Fig. 6.13, there is a comparison of the three different models. Notice that the smallest p-value (significant) is associated with the linear model and supports the assumption that there is a linear relationship between the independent and dependent continuous variables.

To further illustrate these nonlinear regressions, consider the following example. During the development of liquid dosage form where reconstitution is required, the investigator wanted to evaluate the development of degradants over time. One formulation with sucrose and a pH of 6.8 was stored at 60 °C. Table 6.3 gives the results at various time points during a 16-day period. The results in the top of Fig. 6.14 shows that there is a significant linear relationship, $F = 39.91, p = 0.001$, and the coefficient of determination is 0.8887 meaning the straight line accounts for 88.9% of the variability on the y-axis. However, the investigator is concerned because the first two data points and the last data point are below the line, while the others are above the line. Would a curvilinear relationship better fit the data? Using Minitab the investigator tests for other shapes and finds a quadratic relationship produces an $F = 8.64, p = 0.042, R^2 = 96.48\%$; where as a cubic relationship produced the best fit with $F = 51.61, p = 0.004, R^2 = 98.10\%$. All three results are statistically significant. Notice the coefficient of determination is associated with the cubic model is the largest. The lower graph in Fig. 6.14 shows the cubic plot and above it the cubic equation used to calculate the curves. Visually notice the randomness of the residuals being above and below the curved line drawn by the computer.

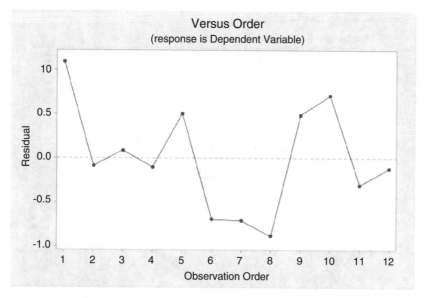

Results of curvilinear options:

Source	DF	SS	F	P
Linear	1	5.88112	14.28	0.004
Quadratic	1	0.50649	1.26	0.290
Cubic	1	0.60917	1.62	0.238

Fig. 6.13 Minitab output and graphic for a residual plot

Table 6.3 Results comparing of time with degradants

Time (days)	Degradants (%)
2	0.1797
3	0.2130
4	0.2912
6	0.3198
8	0.3473
12	0.4002
16	0.4355

6.2.5   Minitab Applications

Minitab can calculate an ANOVA table for linear regression, report the slope and intercept point, graphically plot the data and confidence bars, and evaluate nonlinear relationships. Steps for performing these tests are presented below.

Anova table for linear regression:

Source	DF	SS	MS	F	P
Regression	1	0.0462768	0.0462768	39.91	0.001
Error	5	0.0057975	0.0011595		
Total	6	0.0520743			

S	R-sq	R-sq(adj)
0.0340514	88.87%	86.64%

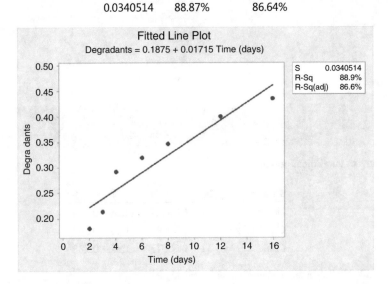

Cubic plot of the sample data:

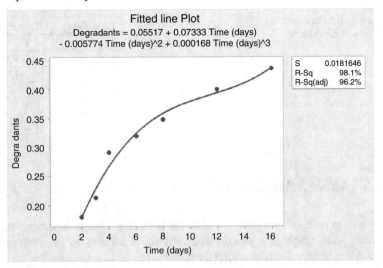

Fig. 6.14 Minitab output and graphics for a comparison of time versus amount of degradants

6.2.5.1   Linear Regression

Procedure	Stats → Regression → Fitted Line Plot
Data input	Select the dependent variable for "Response (Y)" and the independent variable for "Predictor (X)."
Type of regression model	The dialog box defaults to a linear model, but quadratic or cubic can be selected.
Graph	A variety of residual plots are available.
Options	Allow data to be log transformed and to create confidence intervals or predictive intervals. Also, the default type I error is 95% and can be changed. If confidence or prediction bands are required, these can be selected.
Storage	Information can be generated and stored on the worksheet in the next available column(s) for each data point including a variety of residual options.
Report	The ANOVA table is displayed with the F-value and p-value displayed. A graph of the data and line of best fit will be displayed, as well as confidence bands if requested. Additional information on r^2 (expressed as R^2) and pooled standard deviation (S) are presented. Various residual plots will be displayed if requested.
Interpretation	Most important is the p-value in the ANOVA table. If $p < 0.05$, the alternative hypothesis that a linear relationship can be accepted. The value reported for r^2 indicates how tight the data is to the line of best fit.

6.2.5.2   Confidence and Predictive Bands

Procedure	Stats → Regression → Fitted Line Plot
Data Input	Same input as linear regression
Options	Select confidence intervals or predictive intervals to create graphs.
Report	Graph with the data, best fit line, and confidence bands and/or predictive bands.

6.2.5.3   Nonlinear Regression

Procedure	Stats → Regression → Fitted Line Plot
Data input	Select the dependent variable for "Response (Y)" and the independent variable for "Predictor (X)."

Type of regression model	The dialog box defaults to a linear model, select either quadratic or cubic alternatives. For more detailed models use Stats → Regression → Nonlinear Regression (over 25 models are available).
Report	Results will be reported for linear and nonlinear results with associated *p*-values.
Interpretation	Select the model with the smallest associated *p*-value.

6.2.6   Examples

Samples of a drug product are stored in their original containers under normal conditions and sampled periodically to analyze the content of the medication. The results of the 4-year study are presented in Table 6.4. Does a linear relationship exist between the two variables? If such a relation exists, what are the slope and *y*-intercept? Also, based on the study results, can the company be 95% confident that at least 90% of the product (900 mg) will still be available after 4 years? Initial results are presented in Fig. 6.15 and shows that there is a significant linear relationship with $F = 36.30$, $p = 0.004$. The coefficient of determination indicates that the regression line through the sample data (slope = 1.127, *y*-intercept = 995.7 from the graph) accounts for 90.07% of the variability on the *y*-axis. To determine if 90% of the drug is still available after 4 years, the researcher can view the confidence bands in the graph and determine that the confidence interval at 48 months is well above 900 mg. of remaining drug. If the equation in Fig. 6.12 were used to calculate the confidence interval at 48 months, the lower limit would be 927.06 mg, well above the desired amount of drug after 4 years. If the researcher was concerned that a nonlinear fit would be more appropriate, she could use Minitab to also test quadratic and cubic models. The result would be linear $p = 0.004$, quadratic $p = 0.008$ and cubic $p = 0.842$. This shows that the linear model would be best with the smallest type I error of the three models.

For a second example, during the validation of a method, the investigator needs to prove linearity over the range of the test. For a drug product, linearity should be shown with a minimum range of 80–120% (ICH Q2) International Conference

Table 6.4 Results comparing amount of active ingredient measured over time

Time (months)	Assay (mg)
6	995
12	984
18	973
24	960
36	952
48	948

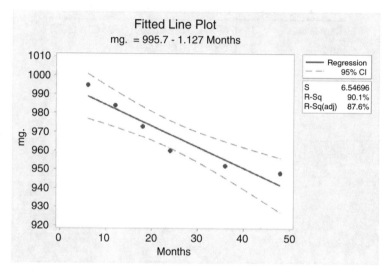

Source	DF	SS	MS	F	P
Regression	1	1555.88	1555.88	36.30	0.004
Error	4	171.45	42.86		
Total	5	1727.33			

S	R-sq	R-sq(adj)
6.54696	90.07%	87.59%

Fitted Line Plot
mg. = 995.7 - 1.127 Months

Fig. 6.15 Minitab output and graphic of linear regression on stability data

on Harmonization (2005). Other in-house SOP standards require that good linearity is reflected by a coefficient of determination equal to or greater than 0.995 and the y-intercept should be no more than 1.5% of the target (zero). The investigator performed the tests within the recommended range, and the results are presented in Table 6.5. Minitab results for the test are seen in Fig. 6.16, and the results are very positive. There is a significant linear relationship ($F = 625.60, p < 0.001$); the coefficient of determination ($r^2 = 0.9952$) is greater than the required 0.995, and the intercept point in the graph is 0.2%, less than the required criteria of 1.5%.

6.3   Multiple Regression

Multiple regression is a logical extension of the concepts illustrated for simple linear regression, where there is more than one predictor variable. Rather than using values for only one predictor or independent variable (to estimate values on a dependent or response variable), with multiple regression, it is possible to evaluate several predictor variables at the same time and look at the response. By using many predictor

Table 6.5 Results of method
validation for linearity

Concentration	Response
0.80	0.79
0.90	0.89
1.00	1.01
1.10	1.08
1.20	1.19

Source	DF	SS	MS	F	P
Regression	1	0.09801	0.0980100	625.60	0.000
Error	3	0.00047	0.0001567		
Total	4	0.09848			

S	R-sq	R-sq(adj)
0.0125167	99.52%	99.36%

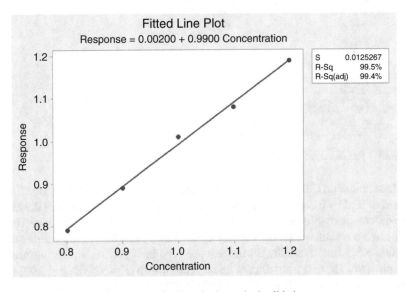

Fig. 6.16 Minitab output and graphic for linearity in method validation

variables (sometime referred to as *exploratory variables*), there is a chance to reduce the error of prediction even further, by accounting for more of the variability, and at the same time, one should be able to increase accuracy of the predictions.

With multiple regression the result is predicting a value for the response variable given a value for each of several corresponding predictor variables. The primary objectives for multiple regression are to (1) determine whether a relationship exists between two or more variables; (2) describe the nature of these relationship, if any exists; and (3) assess the relative importance of each of the various predictor variables and their contribution to the response variable.

The predictor variables can be either continuous or discrete. The calculations for multiple regression are extensive, complex, and fall beyond the scope of this book. Excellent references for additional information on this topic include Zar (pp. 423–437), Snedecor and Cochran (1989), and the Sage University Series (Berry and Feldman 1985; Achen 1982; Schroeder et al. 1986). However, with the use of Minitab and other software packages, these calculations are easily accomplished, and the reported results help determine the importance of each predictor variable.

The results of multiple regression are presented in an ANOVA table similar to the results for an N-way ANOVA (Sect. 5.7.2). The results for each predictor variable and its relationship to the response variable are presented as an F-statistic and corresponding p-value. Interpretation is similar to other tests; if the p-value is less than 0.05, there is a significant relationship between the two variables.

6.3.1 Minitab Applications for Multiple Regression

Procedure	Stats → Regression → Fit Regression Model
Data input	Select the dependent variable for "Responses"; any independent discrete variables for "Categorical predictors" and any independent continuous variables for "Continuous predictors."
Graph	A variety of residual plots are available.
Options	Default type I error is 95% and two-sided test can be changed. Also weighting and transformations are possible.
Storage	Information can be generated and stored on the worksheet in the next available column(s) for each data point including a variety of residual options.
Report	The ANOVA table is displayed with the F-value and p-value displayed.
Interpretation	Most important is the p-values in the ANOVA table. If $p < 0.05$, that factor or interaction significantly influences the outcome on the dependent variable.

6.3.2 Example

An instrument manufacturer is producing a new dissolution tester and wants to try different diameters for the shaft and the width of the paddle. Dimensions are within USP standards (General Chapter 711) USP 42-NF 37 (2019b) with shaft diameters of 10 mm (high) and 9.5 mm (low) and paddle widths of 75 mm (high) and 44 mm (low). The concern is whether either of these dimensions will affect the results of the dissolution test. As a further precaution, three different technicians will set up

Table 6.6 Results of dissolution tests (average for six vessels reported as percent at 20 minutes)

Diameter	High				Low			
Radius	High		Low		High		Low	
Tester	A	B	A	B	A	B	A	B
Technician								
AJ	60.93	60.80	60.77	59.57	60.37	61.23	60.98	60.27
CC	60.72	61.05	60.72	60.03	60.83	61.12	60.82	60.13
FG	61.09	61.58	60.93	61.95	60.68	61.87	61.02	60.53

Source	DF	Adj SS	Adj MS	F-Value	P-Value
Regression	5	13.852	2.77038	1.61	0.162
Technician	2	5.375	2.68750	1.56	0.214
Tester	1	0.095	0.09507	0.06	0.815
Diameter	1	0.020	0.02007	0.01	0.914
Radius	1	8.362	8.36174	4.85	0.029
Error	138	237.707	1.72252		
Total	143	251.559			

Fig. 6.17 Minitab output of multiple regression for a new dissolution apparatus

and run test on two different testers, both of the new design. A summary of the results using the same reference standard for the percent of dissolution at 20 minutes are presented in Table 6.6. Did any of the factors (diameter, radius, technician, or tester) show significant effects on the rate of dissolution? The results of the multiple regression appear in Fig. 6.17. In this case the only significant factor affecting the outcome of the study was the radius of the paddle. The larger and smaller radiuses gave different results when all the factors were considered. A further two-sample t-test comparing all the radius measures and removing the possible confounding variables found the same results ($t = 2.21, p = 0.029$).

6.4   Logistic Regression

Logistic regression is another predictive test where the outcomes are binary discrete results. For example, pass or fail an analysis, where there are only two possible outcomes. Like multiple regression there can be one or more predictor variable, and these predictive variables can be continuous or discrete. In Minitab it is referred to as *binary logistic regression*. Evaluation of the results are also similar to multiple regression; however, in this case the sample statistic is a chi square value (to be discussed in the next section). Regardless, there are associated p-values, and the interpretation is the same; any p-value less than 0.05 is significant.

6.4.1   Minitab Applications for Logistic Regression

Procedure	Stats → Regression → Binary Logistic Regression → Fit Binary Logistic Regression
Data input	Select the dependent variable for "Responses"; any independent discrete variables for "Categorical predictors" and any independent continuous variables for "Continuous predictors."
Graph	A variety of residual plots are available.
Options	Default type I error is 95% and two-sided test can be changed. Also weighting and transformations are possible.
Storage	Information can be generated and stored on the worksheet in the next available column(s) for each data point including a variety of residual options.
Report	The ANOVA table is displayed with the F-value and p-value displayed.
Interpretation	Most important are the p-values in the ANOVA table. If $p < 0.05$, that factor or interaction significantly influences the outcome on the dependent variable.

6.4.2   Example

In the previous example of multiple regression where the instrument manufacturer is testing a new dissolution apparatus, it was decided instead to use a pass/fail criteria rather than percent for determining any significant differences with the variables tested. The grand mean for all the tests was 60.80%, and it was decided a priori that anything 1.5% less than the average would be considered a failure. Thus, results less than 59.3% were considered failures.

The results of the logistic regression based on pass/fail criteria appear in Fig. 6.18. In this case none of the four predictor variables were found to be significant with all the p-values greater than 0.05.

6.5   Chi Square and Related Tests

The chi square tests are used when only discrete variables are involved. In the goodness-of-fit test, there is one discrete variable. For the test of independence, two or more discrete variables are compared: one might be independent (e.g., experimental versus control group), and the other is dependent upon the first (e.g., met goal versus did not meet goal). The chi square test, often referred to as *Pearson's chi square*, evaluates the importance of the difference between what is expected (under

Source	DF	Adj Dev	Adj Mean	Chi-Square	P-Value
Regression	5	7.054	1.4108	7.05	0.217
Technician	2	4.487	2.2436	4.49	0.106
Diameter	1	0.296	0.2956	0.30	0.587
Radius	1	1.182	1.1817	1.18	0.277
Tester	1	1.182	1.1817	1.18	0.277
Error	138	93.410	0.6769		
Total	143	100.464			

Fig. 6.18 Minitab output of logistic regression for a new dissolution apparatus

given conditions) and what is actually observed. When criteria are not met for the chi square test of independence, Fisher's exact test may be used. Pairing of dichotomous outcomes is possible using McNemar's test, and the effects of a third possible confounding variable can be addressed using the Cochran-Mantel-Haenszel test. All of these tests are available on Minitab.

The chi square (χ^2) can best be thought of as a discrepancy statistic. It analyzes the difference between observed values and those values that would normally expect to occur. It is calculated by determining the difference between the frequencies actually observed in a sample data set and the expected frequencies based on probability. Some textbooks classify the chi square and related tests as nonparametric procedures because they are not concerned with distributions around a central point and do not require assumptions of homogeneity or normality.

With the chi square statistics, the frequencies counts in each cell (defined as the level of the dependent discrete variable) are evaluated. The calculation involves squaring the differences between the observed and expected frequencies divided by the expected frequency (Fig. 6.19). The greater the disparity between the observed and expected, the larger the numerator, the larger the χ^2 statistic, and the greater the likelihood that there is a significant relationship.

6.5.1 Chi Square Goodness-of-Fit Test

The goodness-of-fit model can be used for a single discrete dependent variable where certain levels are expected to normally occur. For example, assume different batches of a particular product are compared, and one would expected to see equal results for each batch based on some measure that has discrete outcomes. In this case assume that four different batches a particular drug evaluated for some minor undesirable trait (e.g., a blemish on the tablet coating). Suppose that 200 tablets are randomly sample from each batch and tablets examined for that trait. If the batches are equal, one would expect the same number (frequency count) of failures (blemishes) in each batch. However, the results are Batch A has five, Batch B has two, Batch C has nine, and Batch D has six blemishes. Statistically are these observed

$$\chi^2 = \sum \frac{(f_O - f_E)^2}{f_E} = \sum \frac{(O - E)^2}{E}$$

Where: f_O or O is the number of observations in the sample
f_E or E is the number expected under normal or independent
conditions

Fig. 6.19 Equation for the chi square statistic

differences among the four batches statistically significant or is the variability due to random chance? In this case the hypotheses would be:

H_0 : The samples are selected from the same population
H_1 : The samples are from different populations

If unable to reject the null hypothesis ($p > 0.05$), then the researcher failed to find a significant difference among the four batches. If no difference between batches the expected number of blemishes would be the average for the four batches, which is 5.5 blemishes. The chi square statistic would compare what is expected (5.5) compared to what was observed in each of the four batch samples.

The goodness-of-fit test could also be used to determine if distributions fit proposed models, for example, a binomial or a normal distribution. These model tests are discussed by De Muth (pp. 413–417).

6.5.2   Chi Square Test of Independence

The most common use of the chi square test is to determine if two discrete variables are independent of each other. If not independent, then one variable influences the outcome on the second variable, and there is a relationship between these two discrete variables. Also referred to as the *chi square test for association*, this test is concerned with conditional probability, where the probability for some level of variable A *given* a certain level of variable B is calculated as follows:

$$p(B) \text{ given } A = p(B|A) = \frac{(B \cap A)}{p(A)}$$

If the two discrete variables are independent of each other, then the probability of B for each given level A should be the same regardless of which level of the A variable.

$$p(B_1|A_1) = p(B_1|A_2) = \ldots (B_1|A_k) = p(B_1)$$

Fig. 6.20 Design of the contingency table for a chi square test of independence

Levels of the First Variable

		A_1	A_2	A_3	...	A_K
	B_1				...	
Levels of the Second Variable	B_2				...	
	...	...	...	...	...	
	B_j				...	

A *contingency table* is created where frequency of occurrences are listed for the various levels of each variable. These cells within the contingency table represent frequency counts (not probabilities), and the table is used to determine whether two discrete variables are contingent or dependent on each other. The table has a finite number of mutually exclusive and exhaustive categories in the rows and columns (Fig. 6.20). This bivariate table can be used to predict if two variables are independent of each other or if an association exists. The null hypothesis implies that there is no relationship (complete independence) between the two variables based on probabilities and that each is independent of the other.

$$H_0: \quad P\left(B_1|A_1\right) = P\left(B_1|A_2\right) = P\left(B_1|A_3\right)... = P\left(B_1|A_K\right) = P\left(B_1\right)$$
$$P\left(B_2|A_1\right) = P\left(B_2|A_2\right) = P\left(B_2|A_3\right)... = P\left(B_2|A_K\right) = P\left(B_2\right)$$
$$...$$
$$P\left(B_J|A_1\right) = P\left(B_J|A_2\right) = P\left(B_J|A_3\right)... = P\left(B_J|A_K\right) = P\left(B_J\right)$$

$H_1: \quad H_0$ is false

The chi square test of independence determines if the hypothesis that two variables are related or if the results occur simply by chance. Simpler statements for the previous hypotheses would be:

$H_0:$ Factor B is independent of Factor A

$H_1:$ Factor B is not independent of Factor A

Thus, in the null hypothesis, the probability of B_1 (or B_2 ... or B_M) remains the same regardless of the level of the second variable, A. If one fails to reject H_0, the two variables have no systematic association or relationship. The test statistic in Fig. 6.19 is applied for each cell within the contingency table. The number of degrees of freedom is determined by the number of row minus one times the number of column minus one $((R-1) \times (C-1))$. For a four-by-six contingency table, the number of degrees of freedom would be 15 $((4-1) \times (6-1))$. Critical values for the chi square tests are presented in Table B5 (Appendix B). Determination of significance and rejection of the null hypothesis would be based on a calculated chi square value greater than the critical value in the table or a p-value less than 0.05.

Observed values are easy, simply look at the values in the contingency table. But what are the expected values under complete independence (the null hypothesis)?

The observed value for any cell is the total number of observations for that cell's row times the total number of observations for the cell's column, divided by the total number of observations $((C_{total} \times R_{total})/N)$.

For the chi square test of independence, there are two general rules: (1) there must be at least one observation in every cell (no empty cells or frequency of zero), and (2) the expected value for each cell must be equal to or greater than five. NOTE: expected values less than five, NOT observed values. The chi square formula is theoretically valid only when the expected values are sufficiently large. If these criteria are not met, adjacent rows or columns should be combined so that cells with extremely small values or empty cells are combined to form cells large enough to meet these criteria (see example below).

To meet the criteria if the rows and columns are combined, the smallest possible chi square design is a two-by-two contingency table. For this case only, the "observed minus expected" formula (Fig. 6.19) can be replace by the one presented in Fig. 6.21. Either formula for a two-by-two contingency table will give the same results.

Yates and colleagues argued that the chi square statistics becomes stressed as the number of degrees of freedom decrease (Yates 1934). The most extreme case would be a two-by-two contingency table where there is only one degree of freedom. Yates offers an alternative formula (Fig. 6.21) which creates a slightly more conservative chi square value. Unfortunately, Minitab does not provide this option. One should be mindful that this exists when reading articles/reports that may indicate that a Yate's chi square or a "corrected" chi square was used in the calculations.

6.5.3 Minitab Applications for Chi Square

Minitab has applications for both the goodness-of-fit test and test of independence for chi square. It does not offer the more conservative Yate's correction. Steps for performing these tests are presented below.

Fig. 6.21 Alternative chi square statistics for a two-by-two contingency table

Pearson chi square:

$$\chi^2 = \frac{n(ad - bc)^2}{(a + b)(c + d)(a + c)(b + d)}$$

where:

	A_1	A_2	
B_1	a	b	a+b
B_2	c	d	c+d
	a+c	b+d	N

Yate's corrected chi square:

$$\chi^2_{corrected} = \frac{n(|ad - bc| - .5n)^2}{(a + b)(c + d)(a + c)(b + d)}$$

6.5.3.1   Chi Square Goodness-of-Fit Test

Procedure   Stats ➜ Tables ➜ Chi Square Goodness-of-Fit Test (One variable)

Input   For "Observed counts" enter the frequencies separated by a space. For "Categories" results are listed in one column with each row representing a labeled result for each level of the variable being tested.

Tests   Three options are available: (1) assume equal results for each level on the variable; (2) assign expected proportions for each level of the variable; or (3) assign expected frequency counts for each level of the variable tested.

Graph   By default graphs will be generated displaying bar charts for: (1) the observed and expected values; and (2) how each level contributes to the chi square statistic (which can be further displayed from the largest to smallest contribution to the results).

Results   By default the test statistic displayed.

Report   A table is presented with the observed value, expected values, and expected proportion for each level of the variable and how much each level contributed to the chi square statistic. Also chi square test results are reported with the chi square statistic and associated p-value.

Interpretation   Most important is the p-values for the chi square statistic. If $p < 0.05$, reject the null hypothesis that there are no significant differences among the levels of the variable tested. In the table, the level with the largest "contribution" had the greatest amount of difference between the observed and expected results.

6.5.3.2   Chi Square Test of Independence

Procedure   Stats ➜ Tables ➜ Chi Square Test for Association

Input   Two options are available: (1) select columns from the worksheet indicating which are the "Column" and "Row" variables, or (2) enter the table onto the worksheet with the column variable levels labeled. If the latter option, enter the column names in "Columns containing the table."

Statistics   By default will calculate the chi square test and display counts for each cell, margin sums, and expected values for each cell. Additional information can be selected for raw, standardized, and adjusted residuals and how much each cell contributes to the chi square statistic.

Options   By default will display the row and columns information. Additional information can be requested for a display of missing values.

Report	A table is presented based on the information requested in "Statistics" and "Options." Chi square (Pearson) and likelihood statistics are reported with corresponding p-values. Any potential violations of chi square rules (empty cells or expected values less than five) will be reported below the statistics.
Interpretation	Most important is the p-values for the chi square (Pearson) statistic. If $p < 0.05$, reject the null hypothesis that the column and rows variables are independent of each other and conclude that a significant relationship exists.

6.5.3.3   Alternative Method for the Chi Square Test of Independence

Procedure	Stats ➜ Tables ➜ Cross Tabulation and Chi Square
Input	Same as above but can request a display of row percents, column percents, and total percents. Also, for three-dimensional models, a third variable can be tested as "Layers."
Chi square	Must check "chi square test" to get the inferential statistic. Additional information can be selected for expected cell counts, raw, standardized, and adjusted residuals and how much each cell contributes to the chi square statistic.
Other stats	Other chi square-related tests can be requested including Fisher's exact test, McNemar's test, and Cochran-Mantel-Haenszel test (Sect. 6.5.5).
Options	If data is entered from columns, additional information can be requested.
Report	A table is presented based on the information requested in "Statistics" and "Options." Chi square (Pearson) and likelihood statistics are reported with corresponding p-values. Any potential violations of chi square rules (empty cells or expected values less than five) will be reported below the statistics.
Interpretation	Most important is the p-values for the chi square (Pearson) statistic. If $p < 0.05$, reject the null hypothesis that the column and rows variables are independent of each other and conclude that a significant relationship exists.

6.5.4   Examples

Using the previous example (Sect. 6.5.1), the researcher wanted to know if there is a significant difference in the number of blemishes in the four batches tested. The null hypothesis would be that the proportion of blemish is the same for all four batches. The result of the evaluation is presented in Fig. 6.22 where $\chi^2 = 4.545$. With $p = 0.208$, the researcher failed to reject the null hypothesis and concluded that he

Category	Observed	Test Proportion	Expected	Contribution to Chi-Square
1	5	0.25	5.5	0.04545
2	2	0.25	5.5	2.22727
3	9	0.25	5.5	2.22727
4	6	0.25	5.5	0.04545

N	DF	Chi-Sq	P-Value
22	3	4.54545	0.208

Fig. 6.22 Minitab output for a chi square goodness-of-fit test for four batches of a drug product

was unable to identify a significant difference in the number of blemished among the four batches base on the sample collected. Going back to hypothesis testing the Chap. 4, note that he did not prove the rate of blemishes were same for each batch (the null hypothesis), he simply failed to find a significant difference.

In a second example, a manufacturer of over-the-counter products is evaluating the taste for three formula of a cough suppressant. Volunteers are asked to evaluate the taste based on sweetness using a JAR (just about right) scale. This five-point Likert scale has the following ordinal levels: 1 way too sour, 2 too sour, 3 just about right, 4 too sweet, and 5 way too sweet. The results for the volunteers are reported on Table 6.7. Was there a significant relationship between the three formulas and taste preferences or were the two variables independent of each other? Results of a chi square test of independent show that there was not a significant relationship ($p = 0.061$) between the formulas and taste scores (top of Fig. 6.23). However, Minitab warned that rules were violated because six cells had expected values less than five (expected values in Fig. 6.23). To correct this, the researcher can combine the two lower scores and two higher scores to create a three-by-three contingency table. In this newly created table, all the expected values are greater than five, and the one empty cell (where the observed result was zero) has been removed (lower portion of Fig. 6.23). The new calculation for chi square identifies a significant relationship ($p = 0.040$) and suggests that the formula does effect the responses to the taste test.

6.5.5 Tests Related to Chi Square

Three useful tests related to the chi square test are available on Minitab. All three require that the design be a two-by-two contingency table (two levels for the discrete column variable and two levels for the discrete row variable). The first is *Fisher's exact test*. This test is extremely useful for small samples or where a chi square test of independence may not comply with the required rules: (1) at least one observation in every cell and (2) all expected value for equal to or greater than five. For example, the contingency table has been collapsed to a two-by-two contingency

Table 6.7 Results of taste comparisons for different formulas for a cough suppressant

	Taste scale					
	1	2	3	4	5	Totals
Formula A	0	8	35	6	1	50
Formula B	3	7	32	7	1	50
Formula C	0	2	45	2	1	50
	3	17	112	15	3	150

Initial chi square:

	Chi-Square	DF	P-Value
Pearson	14.929	8	0.061

6 cell(s) with expected counts less than 5.

Expected values:

	1	2	3	4	5
1	1.000	5.667	37.333	5.000	1.000
2	1.000	5.667	37.333	5.000	1.000
3	1.000	5.667	37.333	5.000	1.000

Cell Contents
  Expected count

Revised design to remove expect values < 5:

	Taste Scale			
	Too sour	JAR	Too sweet	
Formula A	8	35	7	50
Formula B	10	32	8	50
Formula C	2	45	3	50
	20	112	18	150

Chi Square for a three-by-three contingency table:

	Chi-Square	DF	P-Value
Pearson	10.015	4	0.040

Fig. 6.23 Minitab output for the analysis of cough suppressant formulas and taste results

table, and it still fails to meet these two requirements. In this case Fisher's exact test is appropriate. The term "exact" implies that there is no intermediate statistic and that the test result is the p-value. The equation for Fisher's exact test is presented in Fig. 6.24 where factorials for the cells and margins for a two-by-two contingency table are used in the equation. However, if there are any more extreme possible outcomes, additional calculations are needed. For example, with a 2×2 contingency table with two in the upper left corner (seen in the example below), additional Fisher's exact tests would need to be calculated (keeping the margins the same,

Fig. 6.24 Equations for   Fisher's Exact Test
tests related to the chi
square test of
independence

$$p = \frac{(a + b)! \times (c + d)! \times (a + c)! \times (b + d)!}{N! \times a! \times b! \times c! \times d!}$$

McNemar's Test

Estimated difference:

$$\hat{\delta} = \frac{b - c}{n}$$

Confidence interval:

$$\delta = \hat{\delta} \pm \left(Z_{\alpha/2} \times \frac{\sqrt{b + c - n\hat{\delta}^2}}{n} + \frac{1}{n} \right)$$

Cochran-Mental-Henzel Test

$$\chi^2_{CMH} = \frac{\left[\sum \frac{a_i d_i - b_i c_i}{N_i} \right]^2}{\sum \left[\frac{(a_i + b_i)(c_i + d_i)(a_i + c_i)(b_i + d_i)}{N_1^2(n_1 - 1)} \right]}$$

where: see contingency table in Figure 6.21 for letters
i = each level of confounding variable

sums for the rows and columns and N) where one and zero are in the upper left corner. These three p-values would be summed and multiplied by two (for a two-tailed test) to calculate a final p-value for Fisher's exact test. The result of the calculation would be the p-value, and any result less than 0.05 would indicate a significant relationship between the row and column variables.

The second test is the *McNemar's test* which is a paired test involving two discrete outcomes. The test involves dichotomous measurements (e.g., pass/fail, yes/no, present/absent) that are paired. The paired responses are constructed into a fourfold or two-by-two contingency table, and outcomes are tallied into the appropriate cell. Measurements can be paired on the same individuals or samples over two different time periods (similar the paired t-test in Chap. 5), and the layout would be to have the initial measure in columns and second measure in the rows of the contingency table. For example, referring to Fig. 6.25, if it were based on a yes/no response over two time periods, those individuals responding "yes" at both time periods would be counted in the upper left corner (cell a), and those answering "no" on both occasions are counted in the lower right corner (cell d). Mixed answers, indicating changes in responses, would be counted in the other two diagonal cells (b and c). If there was absolutely no change over the two time periods, we would expect that 100% of the results would appear in cells a and d. Those falling in cells c and b represent changes between the two measurement periods and are of primary

Fig. 6.25 Sample layout for data for McNemar's test

		Time 1	
		Yes	No
Time 2	Yes	a	b
	No	c	d

interest and evaluated using the formula in Fig. 6.24. The null hypothesis would be that there is no significant change between the two times or characteristics. Thus, a confidence interval that includes zero would result in a failure to reject the null hypothesis. Minitab uses a binomial approach to calculate the p-value, and as with previous tests, the smaller the p-value, the greater the significant relationship between the points tested and the outcomes.

The last test is the *Cochran-Mantel-Haenszel test* which addresses the possibility of a third confounding variable and its effects on the two discrete variables. The test is sometimes referred to as the *Mantel-Haenszel test* or *Mantel-Haenszel-Cochran test* and can be thought of as a three-dimensional chi square test, where a two-by-two contingency table is mandatory for the main factors in the row and columns. However a third, possibly confounding variable, may have as many levels are requited in the research design. This third extraneous factor will have k-levels, and the resultant design would be two-by-two-by-k levels of three discrete variables. The equation presented in Fig. 6.24 compares k different two-by-two contingency tables, one for each of those k-levels. The null hypothesis reflects independence between the row and column variables, correcting for the third extraneous factor. The calculated χ^2_{CMH} is compared to the chi square critical value for one degree of freedom (3.8414). If that value exceeds the critical value or if the p-values are less than 0.05, the third confounding factors have a significant effect on the other two discrete variables. Once again, Minitab uses a slightly different formula with a correction for continuity factor to solving for the CMH test, and the results maybe slightly more conservative (larger p-values) than the equation in Fig. 6.24.

Many other related test of association are available but beyond the scope of this book. Some are available in Minitab, and references for these tests are available (De Muth 2014, pp. 447–475; Goodman 1979; Liebetrau 1983; Zar 2010, pp. 517–567).

6.5.5.1 Minitab Applications

Minitab has applications for Fisher's exact test for small samples, McNemar's test for paired discrete variables and the Cochran-Mantel-Haenszel test for situations where there may be a third confounding variable. The procedure for approaching any of these tests is the same.

Stats → Tables → Cross Tabulation and Chi Square

Any one of the three tests is selected under *Other stats* on the main dialog box. Other procedures are also available which have not been discussed in this book.

Input   Two options are available: (1) select columns from the worksheet
                         indicating which are the "Column" and "Row" variables, or (2)
                         enter the table onto the worksheet with the column variable levels
                         labeled. If the latter option, enter the column names in "Columns
                         containing the table."

   For the Cochran-Mantel-Haenszel test, only the first input method can be used,
and the third confounding variable is added as the *Layers* on the main dialog box.

Chi square   If "Chi square test" is checked will this statistic will also be cal-
                         culated. Additional information can be selected for expected cell
                         counts, raw, standardized, and adjusted residuals and how much
                         each cell contributes to the chi square statistic.
Options   If data is entered from columns, additional information can be
                         requested.
Report   A contingency table will be presented and the requested test sta-
                         tistic with an associated p-value.
Interpretation   If $p < 0.05$, reject the null hypothesis that the column and rows
                         variables are independent of each other and conclude that a sig-
                         nificant relationship exists.

6.5.5.2   Examples

Case 1 In preparing to market an over-the-counter tablet in a new package design,
the manufacturer tests two different blister packs to determine the rates of failure
(separation of the adhesive seal) when stored at various temperatures and varying
degrees of humidity. One thousand tablets in each of two conditions were stored for
3 months, and the number of failures is seen in Table 6.8. Initially a chi square
goodness-of-fit test was performed, but Minitab warned of violation of rules because
two cells have expected values less than five (Fig. 6.26). Therefore, the data was
reanalyzed using Fisher's exact test, and these results are also shown in Fig. 6.26. In
both cases, the researcher failed to find a significant relationship between the type
of blister pack and failures because the p-values were greater than 0.05.

Case 2 Immediately after training on a new analytical method, technicians were
asked their preference between the "new" method and a previously used "old"
method for analysis. Six months later, after the technicians had experience with the

Table 6.8   Number of blister pack failures at different storage conditions

	40° 50% relative humidity	60° 50% relative humidity
Blister pack A	2	5
Blister pack B	8	5

Fig. 6.26 Minitab output for blister pack failures at two different storage conditions

Chi square:

	Chi-Square	DF	P-Value
Pearson	1.978	1	0.160

2 cell(s) with expected counts less than 5.

Fisher's exact test:

P-Value
0.349845

Table 6.9 Results of surveys before and after experience with a new analytical method

		Preferred method before experience		
		New	Old	
Preferred method After experience	New	8	13	21
	Old	2	7	9
		10	20	30

new method, they were resurveyed a second time with respect to their preference. The results of the two surveys are presented in Table 6.9. Did experience with the new method significantly change their preferences? Because it was a paired study (before and after) survey and the results were discrete responses (preferred new or old) method, the surveyor decided to perform a McNemar's test on the data. The results of the analysis are seen in Fig. 6.27. There appears to be a significant switch in the preferences of the analysts toward the new method both with a confidence interval that does not include zero and with a significant p-value of 0.007. In this case 61.9% switch from the old method to the new method, but only 22.2% switch back from preferring the new method to favoring the old method in the final survey.

Case 3 An instrument manufacturer ran a series of disintegration tests to compare the pass/fail rate of a new piece of equipment at two extreme temperatures and failed to find a significant difference ($p = 0.182$). The manufacturer decided to also evaluate the influence of cycle time for the basket to the two extremes in the USP (General Chapter 701) USP 42-NF 37 (2019a). The test was redesigned to collect results at two cycle times (29 and 32 cycles/minute), defined as fast and slow. The results of the study are presented in Table 6.10. In this case there is a three-dimensional model with discrete two-level dependent variables (pass/fail the test) and two independent variables (temperature extremes and cycle time extremes). The manufacturer decided the most appropriate test would be the Cochran-Mantel-Haenszel test, and the results appear in Fig. 6.28. Fortunately for the manufacturer, the results were not confounded by the two extremes in cycle per minute ($p = 0.288$).

Fig. 6.27 Minitab output
for a McNemar's test on
analysts' preferences for
two different methods and
two different times

	New	Old	All
New	8	13	21
	26.67	43.33	70.00
Old	2	7	9
	6.67	23.33	30.00
All	10	20	30
	33.33	66.67	100.00

Cell Contents
Count
% of Total

Estimated Difference	95% CI	P
0.367	(0.117, 0.616)	0.007

Difference = p (After = New) - p (Before = New)

Table 6.10 Results of different number of cycles and temperatures on disintegration results

Cycles/minute	Temperature (°C)	Test results		Totals
		Pass	Fail	
Fast (32)	39	57	3	60
	35	56	4	60
	Totals	113	7	120
Slow (29)	39	58	2	60
	35	54	6	60
	Totals	112	8	120

Chi Square comparing temperature and failure rate:

	Chi-Square	DF	P-Value
Pearson	1.778	1	0.182

CMH test for confounding by cycles/minute:

Common Odds Ratio	CMH Statistic	DF	P-Value
0.479167	1.12863	1	0.288068

Results for all 2x2 tables

Fig. 6.28 Minitab output for CMH evaluation of different number of cycles and temperatures on disintegration results

References

Achen CH (1982) Interpreting and using regression (paper 29). In: Sage University series on quantitative applications in the social sciences. Sage Publications, Newbury Park

Anscombe FJ (1973) Graphs in statistical analysis. Am Stat 27:17–27

Berry WD, Feldman S (1985) Multiple regression in practice (paper 50). In: Sage University series on quantitative applications in the social sciences. Sage Publications, Newbury Park

De Muth JE (2014) Basic statistics and pharmaceutical statistical applications, 3rd edn. CRC Press, Boca Raton

Goodman LA (1979) Measures of association for cross classifications. Springer, New York

Guilford JP (1956) Fundamental statistics in psychology and education. McGraw-Hill, New York, p 145

International Conference on Harmonization (2005) Q2(R1) validation of analytical procedures: text and methodology

Kelly WD, Ratliff TA, Nenadic C (1992) Basic statistics for laboratories. Wiley, Hoboken, p 93

Liebetrau AM (1983) Measures of association. Sage Publications, Newbury Park

Schroeder LD et al (1986) Understanding regression analysis: an introductory guide (paper 57). In: Sage University Series on quantitative applications in the social sciences. Sage Publications, Newbury Park

Snedecor GW, Cochran WG (1989) Statistical methods, 8th edn. Iowa State University Press, Ames, pp 333–365

Spearman CE (1904) The proof and measurement of association between two things. Am J Psychol 15:72–101

USP 42-NF 37 (2019a) General chapter <701> disintegration. US Pharmacopeial Convention, Rockville

USP 42-NF 37 (2019b) General chapter <711> dissolution. US Pharmacopeial Convention, Rockville

Yates F (1934) Contingency tables involving small numbers and the χ^2 test. J Royal Stat Soc 1:217–235

Zar JH (2010) Biostatistical analysis, 5th edn. Prentice-Hall, Upper Saddle River

Chapter 7
Tests to Identify Similarities

Abstract The inferential tests presented in Chap. 5 are concerned with identifying significant differences. If the results are not significant, the researcher fails to reject the hypothesis under test, but could never prove that the hypothesis under test was correct. This chapter looks at tests that would allow the researcher to prove that the levels of an independent variable are equivalent within a prescribed amount of similarity. As with previous test, there is a degree of confidence associated with interpreting the equivalence being analyzed. Tests of equivalence present in this chapter include confidence intervals and ratio methods associated with t-tests for the one-sample, two-sample, and paired situations. Minitab applications for evaluating the level of equivalence are presented. Also the SUPAC guidance for determining the equivalence of dissolution curves for immediate-release solid oral dosage forms is discussed.

Keywords Bioequivalence · Confidence intervals · Equivalence · Equivalence limits · SUPAC

Up to this point, the statistical tests discussed have been concerned with null hypotheses stating no difference (e.g., H_0: $\mu_1 = \mu_2$) or no relationship (e.g., H_0: $r = 0$). These tests were designed to identify significant differences and by rejecting the null hypothesis, proving the alternative hypothesis, and proving inequality (H_1: $\mu_1 \neq \mu_2$). As discussed in Sect. 4.2.2, when finding a result that is not statistically significant, we do not prove the null hypothesis is correct, but simply fail to reject it. The analogy was presented of jurisprudence where the jury will render a verdict of "not guilty," but never "innocent" if they failed to prove the accused guilty beyond a reasonable doubt. Similarly, if our data fails to show that a statistically significant difference exists, we do not prove equivalency. But what if the researcher wants to show equality or at least similarity with a certain degree of confidence? In these cases, tests of equivalence become important.

© American Association of Pharmaceutical Scientists 2019
J. E. De Muth, *Practical Statistics for Pharmaceutical Analysis*, AAPS
Advances in the Pharmaceutical Sciences Series 40,
https://doi.org/10.1007/978-3-030-33989-0_7

7.1   One-Sample Test for Equivalence

The one-sample test for equivalence is similar to the one-sample t-test where a confidence interval is created. This univariate test will estimate, with a certain amount of confidence the true population mean. The result would be presented as:

lower equivalence limit < population mean < upper equivalence limit

Assume that there is a target for the label claim of a drug product (e.g., 100%). The desired result would be confidence interval where 100% is a possible outcome, meaning it would fall between the upper and lower equivalence limits. In this case there are actually two null hypotheses, where results are smaller than the lower equivalence limit (LEL) or larger than the upper equivalence limit (UEL):

$$H_{01} : \quad \text{Difference} \leq \text{LEL}$$
$$H_{02} : \quad \text{Difference} \geq \text{UEL}$$

The alternative hypotheses would be:

$$H_{01} : \quad \text{Difference} > \text{LEL}$$
$$H_{02} : \quad \text{Difference} < \text{UEL}$$

If both null hypotheses are rejected, then equivalence is proven:

$$\text{LEL} < \text{Difference} < \text{UEL}$$

with $1-\alpha$ level of confidence. When trying to prove equivalence to the target value, note that each test is conducted a one-tailed test with $\alpha = 0.05$. Therefore, the value of Table B2 in Appendix B would be under column $t_{0.95}$, instead of the two-tailed value in column $t_{0.975}$. The total type I error would be $p = 0.10$ or 10% chance of being wrong with the decision of equivalence (5% for each hypothesis).

Another approach would be to use a test of equivalence to show that the sample results are within a certain percent of the target. For example, based on sample results, it is estimated that the entire batch of a drug is within 1.5% of the target goal of 100%. The formulas for a one-sample equivalence test are presented in Fig. 7.1. Continuing with this example, the upper and lower limits would be −1.5 and +1.5, respectively. But what if one were measuring something other than percent or a 100% target? For example, tablet breaking force and the goal was to be within 5% of a target of 15.0 kiloponds (kp). In this case the desired equivalence limits would be 15 kp, the lower limit −0.75 kp, and the upper limit +0.75 kp (0.05 × 15 kp). Minitab has several tests for equivalence, including the one-sample case.

Difference: $D = \bar{X} - \text{Target}$

Lower equivalency limit: δ_1 expressed as % target
Upper equivalency limit: δ_2 expressed as % target

Degrees of freedom: $n-1$

Standard error of the difference (SE) : $SE = \dfrac{s}{\sqrt{n}}$

Reliability coefficient $(100(1-\alpha))$%: t-values of Table B2 for $t_{0.95}$

Two one-sided tests
   Confidence Intervals:
$$D_L = D - t_{1-\alpha,v} \times SE$$
$$D_U = D + t_{1-\alpha,v} \times SE$$

  Ratio t-statistic:
$$t_1 = \frac{D - \delta_1}{SE} \qquad t_2 = \frac{D - \delta_2}{SE}$$

Fig. 7.1 Equations for a one-sample equivalence test

7.1.1   Minitab Application

Procedure	Stat ➔ Equivalence tests ➔ 1-Sample
Data input	Two options are available: (1) select a column from the worksheet ("Sample"), or (2) enter summary statistics that includes the sample size, mean, and standard deviation.
Target	Enter the anticipated mean under equivalence.
Alternative hypothesis	By default a two-tailed confidence interval is created. Other options are available. Entries must be made for predetermined lower and upper limits.
Options	By default for a 95% confidence interval for a two-tailed test, it can be changed to a two one-tailed test by checking "Use (1-2 alpha) × 100% confidence."
Graphs	By default an equivalence plot will be displayed. If data is selected from a column, additional plots for a histogram, individual value plot (dot plot), or box plot are available.
Results	By default method, descriptive statistics, difference, and test results are provided. Any of these can be removed.
Report	Descriptive statistics and a confidence interval for possible population are presented. If lower and upper limits are provided, the p-values will be given for both limits. A default equivalence graph and other possible graphics may be displayed.

Interpretation   For equivalence, visually the confidence interval should fall
                     within the limits on the equivalence graph. For numerical equiva-
                     lence the p-values for the upper and lower limits should be less
                     than 0.05.

7.1.2 Example

Six samples are randomly selected from a production batch of a product and tested
to determine if the results are within 1.5% of the label claimed. The sample results
are 99.6, 100.2, 98.3, 99.9, 100.4, and 100.7%, with a mean of 99.85%.

Doing a one-sample t-test, the confidence interval for the true population mean
would fall somewhere between 98.96% and 100.74%. The goal of 100% is a possible outcome, but with 95% confidence is the sample (which represents the entire
batch) equivalent within 1.5% of the target? In this case the upper limit (UEL)
would be +1.5% and lower limit (LEL) would be −1.5%. The Minitab output is
presented in Fig. 7.2. This output provides the results as both a confidence interval
and t-ratio with associated p-value. The t-ratio gives two t-statistics and two p-
values: one for the probability of exceeding the LEL and one for the probability of
exceeding the UEL. In this example of percent label claim, there is a significant
equivalence with the 95% confidence interval between −0.85% and +0.55%, with
$t_{lower} = 3.89$ and $t_{upper} = -4.75$ and with type I errors of 0.006 and 0.003, respectively.
In either case the sample, representing the population, falls within the limits and can
be considered as equivalent to the target of 100%.

Confidence interval:

Difference	SE	95% CI for Equivalence	Equivalence Interval
-0.15000	0.34713	(-0.849486, 0.549486)	(-1.5, 1.5)

CI is within the equivalence interval. Can claim equivalence.

Ratio method:

Null hypothesis:	Difference ≤ -1.5 or Difference ≥ 1.5
Alternative hypothesis:	-1.5 < Difference < 1.5
α level:	0.05

Null Hypothesis	DF	t-value	P-Value
Difference ≤ -1.5	5	3.8890	0.006
Difference ≥ 1.5	5	-4.7532	0.003

The greater of the two P-Values is 0.006. Can claim equivalence.

Fig. 7.2 Minitab output for a one-sample equivalence test

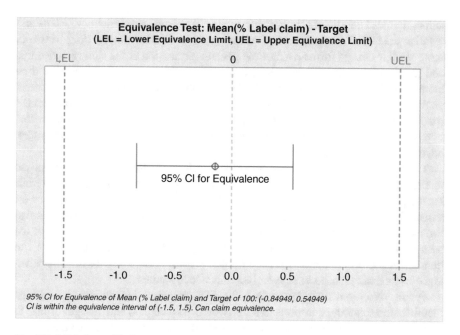

Fig. 7.3 Minitab graphic for a one-sample equivalence test

By default, Minitab provides a graphic representation of the results. Here the confidence interval is well within the lower and upper acceptance limits (Fig. 7.3).

7.2   Two-Sample Test for Equivalence

Much of two-sample equivalency testing relates to bioequivalence testing. When pharmaceutical manufacturers and regulatory agencies began studying the bioequivalence of drug products, the general approach was to use a simple two-sample t-test or analysis of variance to evaluate plasma concentration-time curves (e.g., Cmax, Tmax, AUC). Since these traditional statistical tests were designed to demonstrate differences rather than similarities, they were incorrectly used to interpret the early bioequivalence studies. In the 1970s researchers began to note that traditional hypothesis tests were not appropriate for evaluating bioequivalence (Metzler 1974).

7.2.1   Bioequivalence Testing

In order for an oral or injectable product to be effective, it must reach the site of action in a concentration large enough to exert its effect. Bioavailability indicates the rate and/or amount of active drug ingredient that is absorbed from the product and available at the site of action. *Remington: The Science and Practice of Pharmacy* (Malinowski 2000) defines bioequivalence as an indication "that a drug in two or more similar dosage forms reaches the general circulation at the same relative rate and the same relative extent." Thus, two drug products are bioequivalent if their bioavailabilities are the same and may be used interchangeably for the same therapeutic effect. In contrast to previous tests that attempted to fail to prove differences (two-sample t-tests), the objective of most bioequivalence statistics is to show that two dosage forms are the same or at least close enough to be considered similar, beyond a reasonable doubt.

There are three situations requiring bioequivalence testing: (a) when a proposed marketed dosage form differs significantly from that used in the major clinical trials for the product; (b) when there are major changes in the manufacturing process for a marketed product; and (c) when a new generic product is compared to the innovator's marketed product (Benet and Goyan 1995). Regulatory agencies allow the assumption of safety and effectiveness if the pharmaceutical manufacturers can demonstrate bioequivalence with their product formulations. Two-sample equivalence testing can be used to support bioequivalence studies.

For in vivo bioavailability studies, the FDA requires that the research design identifies the scientific questions to be answered, the drugs(s) and dosage form(s) to be tested, the analytical methods used to assess the outcomes of treatment, and benefit and risk considerations involving human testing (21 Code of Federal Regulations, 320.25(b)). Two types of study designs are generally used for comparing the bioavailability parameters for drugs: (1) the parallel group design or (2) crossover design. Both of which will be discussed below.

7.2.2   Two-Sample Equivalence in a Parallel Study Design

The parallel study design can be accomplished by either a two-sample equivalence test or a paired test for equivalence (Sect. 7.3). In the parallel study design (Fig. 7.4), volunteers are assigned to one of two similar groups, and each group receives only one treatment: in the case of a bioequivalence study, either the test drug or the reference standard, while for analytical testing one of the two test methods. In order to

Fig. 7.4 Parallel study design involving two groups

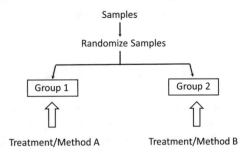

| Difference: | $D = \bar{X}_T - \bar{X}_R$ |

| Lower equivalency limit: | δ_1 expressed as % target |
| Upper equivalency limit: | δ_2 expressed as % target |

| Degrees of freedom: | $n1 + n2 - 2$ |

Standard error of the difference (SE): $\sqrt{\dfrac{2S_p^2}{n}}$ for equal cell sizes

$\sqrt{\dfrac{S_p^2}{n_1} + \dfrac{S_p^2}{n_2}}$ for unequal sizes

Reliability coefficient $(100(1-\alpha))$%: t-values of Table B2 for $t_{0.95}$

Two one-sided tests

Confidence Intervals:
$$D_l = D - t_{1-\alpha,v} \times SE$$
$$D_U = D + t_{1-\alpha,v} \times SE$$

Ratio t-statistic:
$$t_1 = \frac{D - \delta_1}{SE} \qquad t_2 = \frac{D - \delta_2}{SE}$$

Fig. 7.5 Equations for a two-sample equivalence test

establish similar groups, volunteers or samples are randomly assigned to one of the two groups. Because of random assignment to the two treatment levels (groups), it is assumed that each set of volunteers/samples is identical to the other. Therefore, any differences in the bioavailability or analytical measures are attributable to the treatment received.

To overcome the disadvantage of failing to reject the null hypothesis with a two-sample t-test, the two-sample equivalence was developed to identify the similarity or equivalence between two methods or treatments. The two-sample equivalence test is also referred to as the two one-sided t-test (TOST) by the FDA (Schuirmann 1987). It involved testing two hypotheses comparing the treatment level against the reference standard:

$$H_{01}: \quad \mu_T - \mu_R \leq LEL$$
$$H_{02}: \quad \mu_T - \mu_R \geq UEL$$

If both null hypotheses are rejected, then the alternative hypotheses are proven that the results fall within the equivalence limits (LEL < difference < UEL):

$$H_{11}: \quad \mu_T - \mu_R > LEL$$
$$H_{12}: \quad \mu_T - \mu_R < UEL$$

The equations for calculating the two-sample equivalence test are presented in Fig. 7.5. Similar to the one-sample case, determination of equivalence is either (1)

t-values that are significant ($p < 0.05$) or (2) a confidence interval that falls within the lower equivalence limit and upper equivalence limit.

7.2.3 Minitab Application

Procedure	Stat → Equivalence tests → 2-Sample
Data input	Three options are available: (1) sample data can be in a single column ("Samples" is the dependent variable and "Samples ID" is the independent variable); (2) data are on different columns from the worksheet (select "Test sample" and "Reference sample"); or (3) enter summary statistics that include the sample size, mean, and standard deviation for both the test and reference samples.
Hypothesis	Select either differences or ratio methods.
Alternative hypothesis	By default a two-tailed confidence interval is created. Other options are available. Entries must be made for predetermined lower and upper limits.
Options	By default for a 95% confidence interval for a two-tailed test, it can be changed to a two one-tailed test by checking "Use (1-2 alpha) × 100% confidence." Also, by default, the Satterthwaite correction (Sect. 5.3.1) will be applied unless the "Assume equal variances" is checked.
Graphs	By default an equivalence plot will be displayed. For results other than summary data, additional plots for a histogram, individual value plot (dot plot), or box plot are available.
Results	By default method, descriptive statistics, difference, and test results are provided. Any of these can be removed.
Report	Descriptive statistics and a confidence interval for possible population. If lower and upper limits are provided, p-values will be given for both limits. A default equivalence graph and other possible graphics may be displayed.
Interpretation	For equivalence, visually the confidence interval should fall within the limits on the equivalence graph. For numerical equivalence the p-values for the upper and lower limits should be less than 0.05.

7.2.4   *Example*

In the example presented previously (Sect. 5.3.5), samples are taken from a specific batch of drug and randomly divided into two groups of tablets. One group is assayed by the manufacturer's own quality control laboratories. The second group of tablets is sent to a contract laboratory for identical analysis. Performing a two-sample t-test on the sample data in Table 5.2, the investigator failed to find a significant difference between the two laboratories ($t = 1.29$, $p = 0.229$). But are their results equivalent? The investigator would like the results to be within 2% of each other. Since this is based on 100% label claim, the lower limit is −2% and the upper limit is +2%. The results of the analysis are present in Fig. 7.6. In this case the researcher failed for find equivalence within 2%. The confidence interval was from −2.483 to +0.383, the former exceeding the −2% limit. The t-ratio also failed to reject the hypothesis for the lower limit with $t = 1.21$ and $p = 0.128$.

Like the one-sample case, by default Minitab will create a graphic presentation on the results. Here the confidence interval is well within the upper acceptance limits (Fig. 7.7). However, visually the lower end of the confidence interval exceeds the lower equivalence limit, supporting the failure to show equivalence.

7.3   Paired Test for Equivalence

The paired test for equivalence is a compliment to the paired t-test, where no significant differences were identified between the two levels being tested. This test may have limited usefulness since most analytical tests are destructive and a sample could not be tested under two different conditions or methods. But it could be used for comparing the analytical results between the new and senior scientist, seen in the example below. The null hypothesis could not be rejected using a paired t-test ($t = 1.38$, $p = 0.200$). The result was a failure to identify a difference between the two individuals. But it was not possible to prove null hypothesis that $\mu_d = 0$. The paired test for equivalency could be used to determine if their results were within 1, 2, or 3% of each other if performing similar tests over time.

Like the two-sample equivalence test, the paired equivalence test involved testing two hypotheses comparing the treatment level against the reference standard:

$$H_{01} : \quad \mu_d \leq \text{LEL}$$
$$H_{02} : \quad \mu_d \geq \text{UEL}$$

If both null hypotheses are rejected, then the alternative hypotheses are proven that the results fall within the equivalence limits (LEL < difference < UEL):

$$H_{11} : \quad \mu_d > \text{LEL}$$
$$H_{12} : \quad \mu_d < \text{UEL}$$

Confidence interval:

Difference	SE	95% CI for Equivalence	Equivalence Interval
-1.0500	0.78191	(-2.48334, 0.383336)	(-2, 2)

CI is not within the equivalence interval. Cannot claim equivalence.

Ratio method:

Null hypothesis:	Difference $\leq$ -2 or Difference $\geq$ 2
Alternative hypothesis:	-2 < Difference < 2
α level:	0.05

Null Hypothesis	DF	*t*-value	P-Value
Difference $\leq$ -2	9	1.2150	0.128
Difference $\geq$ 2	9	-3.9007	0.002

The greater of the two P-Values is 0.128. Cannot claim equivalence.

Fig. 7.6 Minitab output for a two-sample equivalence test

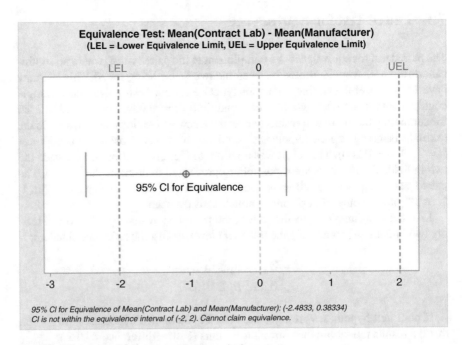

Fig. 7.7 Minitab graphic for a two-sample equivalence test

Difference: $D = \bar{X}_d$ (mean difference for pairs)

Lower equivalency limit: δ_1 expressed as % target
Upper equivalency limit: δ_2 expressed as % target

Degrees of freedom: n-1 (n = number of pairs)

Standard error of the difference (SE) : $SE = \frac{S_d}{\sqrt{n}}$

                                               (S_d= standard deviation for the
                                               difference between pairs)

Reliability coefficient $(100(1-\alpha))$%: t-values of Table B2 for $t_{0.95}$

Two one-sided tests

Confidence Intervals:
$$D_L = D - t_{1-\alpha,v} \times SE$$
$$D_U = D + t_{1-\alpha,v} \times SE$$

Ratio t-statistic:
$$t_1 = \frac{D - \delta_1}{SE} \qquad t_2 = \frac{D - \delta_2}{SE}$$

Fig. 7.8 Equations for a paired equivalence test

The calculations for the paired equivalence test are similar to the one-sample case, except in this case n is the number of pairs, not the total number of data points in the sample (Fig. 7.8). Also, similar to the one-sample and two-sample cases, the Minitab paired equivalence test will give the t-ratio results as well as a confidence interval. Determination of equivalence is either (1) t-values that are significant ($p < 0.05$) or (2) confidence intervals within the lower equivalence limit and the upper equivalence limit.

7.3.1 Minitab Applications

Procedure	Stat → Equivalence tests → Paired.
Data input	Select columns from the worksheet for "Test sample" and "Reference sample."
Hypothesis	Select either differences or ratio methods. A log transformation option also is available.
Alternative hypothesis	By default a two-tailed confidence interval is created. Other options are available. Entries must be made for predetermined lower and upper limits.
Options	By default for a 95% confidence interval for a two-tailed test, it can be changed to a two one-tailed test by checking "Use (1-2 alpha) × 100% confidence."

Graphs	By default an equivalence plot will be displayed. Additional plots for a histogram, individual value plot (dot plot), or box plot are available.
Results	By default method, descriptive statistics, difference, and test results are provided. Any of these can be removed.
Report	Descriptive statistics and a confidence interval for possible population. If lower and upper limits are provided, p-values will be given for both limits. A default equivalence graph and other possible graphics may be displayed.
Interpretation	For equivalence, visually the confidence interval should fall within the limits on the equivalence graph. For numerical equivalence the p-values for the upper and lower difference limits should be less than 0.05.

7.3.2   Example

Following training on content uniformity testing, comparisons are made between the analytical results of the newly trained chemist with those of a senior chemist. Samples of four different drugs (compressed tablets) and different batches of those drugs are selected and assayed by both individuals. In this case each drug/batch is paired for assay by the two chemists. These results are presented in Table 7.1.

As noted previously, if the two scientists are compared using a paired t-test, the results would fail to find a significant difference between the two scientists ($t = 1.38$, $p = 0.200$). However, could their results be considered equivalent within 2% of each

Table 7.1 Results for the two scientists testing the same samples

Sample drug, batch	New chemist	Senior chemist
A,42	99.8	100.4
A,43	99.6	100.2
A,44	101.5	101.3
B,96	99.5	100.6
B,97	99.2	99.4
C,112	100.8	101.5
C,113	98.7	97.9
D,21	100.1	99.9
D,22	99.0	99.3
D,23	99.1	99.2

Confidence interval:

Difference	StDev	SE	95% CI for Equivalence	Equivalence Interval
-0.24000	0.54813	0.17333	(-0.557740, 0.0777396)	(-2, 2)

CI is within the equivalence interval. Can claim equivalence.

Ratio method:

Null hypothesis:	Difference ≤ -2 or Difference ≥ 2
Alternative hypothesis:	-2 < Difference < 2
α level:	0.05

Null Hypothesis	DF	t-value	P-Value
Difference ≤ -2	9	10.154	0.000
Difference ≥ 2	9	-12.923	0.000

The greater of the two P-Values is 0.000. Can claim equivalence.

Fig. 7.9 Minitab output for a paired equivalence test

other? Since this is based on 100% label claim, the lower limit is −2% and the upper limit is +2%. The results of the analysis are present in Fig. 7.9. In this case, equivalence can be proven with the confidence interval falling well within the ±2% range (−0.557 to +0.078). The t-ratios also are significant rejecting both null hypotheses (greater than the upper limit and less than the lower limit) with $p < 0.001$. Like the one-sample equivalence tests and two-sample test, by default Minitab produces a graphic representation on the results. Here the confidence interval is well within the lower and upper acceptance limits (Fig. 7.10). Visually the confidence interval is inside the lower equivalence and upper equivalence limits, supporting the numerical statistical results.

7.4 Crossover Equivalence Test

Minitab can evaluate equivalence for a two-by-two crossover design, for example, in a clinical trial where each volunteer is evaluated under two different conditions (Fig. 7.11). Clinically a two-by-two crossover study could be used to determine if a test drug is equivalent to that of a reference drug where a volunteer serves as his/her own control in the study. In biostatistics it can be used for per-

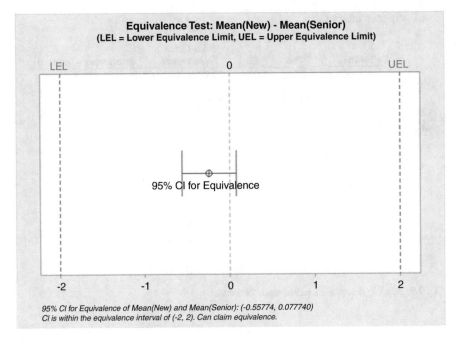

Fig. 7.10 Minitab graphic for a paired equivalence test

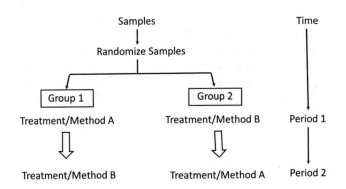

Fig. 7.11 Two-period crossover study design for two groups

forming superiority tests and inferiority tests, to evaluate whether the mean of a new drug is greater than or less than the mean of the commercially available drug. This is beyond the scope of this book because most analytical procedures are destructive and samples would unlikely be crossed over to a second test method. Information regarding this test and Minitab support is available online (https://

support.minitab.com/en-us/minitab/18/help-and-how-to/statistics/equivalence-tests/how-to/equivalence-test-for-a-2x2-crossover-design/before-you-start/overview/).

7.5   Tests for Similarity of Dissolution Profiles

Dissolution tests provide an in vitro method to determine if products produced by various manufacturers or various batches from the same manufacturer are in compliance with compendia or regulatory requirements. Dissolution profiles can be used to compare multiple batches, different manufacturers, or different production sites to determine if the products are similar with respect to percent of drug dissolved over given periods of time. An assumption is that the rate of dissolution and availability will correlate to absorption in the gut and eventually similar effects at the site of action. This assumption can be significantly enhanced if manufacturers can establish an in vivo-in vitro correlation between their dissolution measures and bioavailability outcomes (FDA 1997).

To answer the question of equivalency in dissolution profiles, the FDA has proposed a guidance for manufacturers issued as "Scale-Up and Post-Approval Changes for Immediate Release Solid Oral Dosage Forms" (SUPAC-IR). This guidance is designed to provide recommendations for manufacturers submitting new drug applications, abbreviated new drug applications and abbreviated antibiotic applications to change the process, and equipment or production sites following approval of their previous drug submissions (Federal Register 1995). Previous evaluations had involved single-point dissolution tests. The SUPAC-IR guidance can assist manufacturers with changes associated with (1) scale-up procedures; (2) site changes in the manufacturing facilities; (3) equipment or process changes; and (4) changes in components or composition of the finished dosage form. Under SUPAC-IR there are two factors that can be calculated: (1) a difference factor (f_1) and (2) a similarity factor (f_2). The equations for these factors are presented in Fig. 7.12. The guidance for equivalency is that the f_1-value should be close to 0 (generally values less than 15) and the f_2-value should be close to 100, with values greater than 50 ensuring equivalency. If the two dissolution profiles are exactly the same (one laying exactly over the second), the f_2 value will be 100. As the f_2-value gets smaller, there is a greater difference between the two profiles. An f_2 of 50 represents an approximate 10% difference. Thus, the SUPAC-IR guidance requires a calculated f_2-value between 50 and 100 for equivalence.

Several criteria must be met in order to apply the f_1 and f_2 calculations (FDA): (1) test and reference batches should be tested under exactly the same conditions, including the same time points; (2) only one time point should be considered after 85% dissolution for both batches; and (3) the relative standard deviation at the ear-

Difference factor:

$$f_1 = \frac{\sum |R_t - T_t|}{\sum R_t} \times 100$$

Similarity factor*:

$$f_2 = 50 \log \left[\frac{1}{\sqrt{1 + \frac{1}{n}\sum(R_t - T_t)^2}} \times 100 \right]$$

Where n is the number of time points in the dissolution profile, R_t is the percent dissolved for the reference standard at each time period, T_t is percent dissolved for the test product at the same time period, and log is the logarithm base 10.

* formula was modified for easier readability, originally

$$f_2 = 50 \log \left\{ \left[1 + \frac{1}{n}\sum(R_t - T_t)^2 \right]^{-0.5} \times 100 \right\}$$

Fig. 7.12 Equations for SUPAC-IR

lier time points should be no more than 20% and at other time points should be no more than 10% (Shah et al. 1998).

Unfortunately, Minitab does not include SUPAC-IR applications.

7.5.1 Examples

Two dissolution profiles comparing two pieces of equipment are presented in Table 7.2. Are they similar based on the SUPAC-IR guidance? The study meets the conditions stated above and profiles are presented in Fig. 7.13. The resultant f_2-value is 63.7, thus showing equivalence between the two dissolution profiles.

In a second example, the production of a certain product in two different countries (A and B) is compared to the manufacturer's reference profile for the original production site (standard). Dissolution data is presented in Table 7.3 and Fig. 7.14. Visually it appears that site B has a profile closer to the reference standard, but do both of the foreign facilities meet the SUPAC-IR guidelines for similarity? In this case the calculated f_2-value for site A and the reference standard is 52.9, and the f_2-value for site B and the reference standard is 65.6. Both sites are values larger than 50 and would meet the SUPAC-IR guidance for equivalence.

Table 7.2 Dissolution results (%) comparing two types of equipment

Time (minutes)	15	30	45	60
	Batch produced using original equipment (reference)			
Mean	63.48	76.32	82.85	87.87
SD	7.74	6.79	6.29	5.61
RSD	13.71	8.91	7.59	6.38
	Batch produced using newer equipment (treatment)			
Mean	73.05	79.92	84.32	89.42
SD	10.13	6.33	4.35	3.83
RSD	13.87	7.91	5.16	4.28

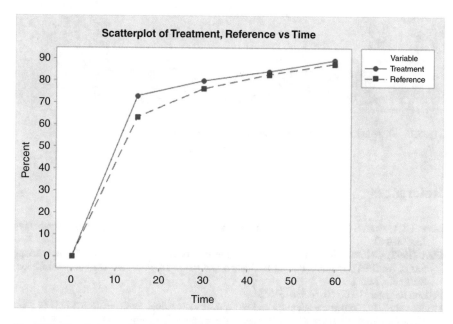

Fig. 7.13 Dissolution profiles for two types of equipment

Table 7.3 Dissolution data (%) comparing two sites to a reference standard

Time (minutes)	Country A	Country B	Standard
15	57.3	54.1	49.8
30	66.4	67.7	70.8
45	71.9	75.4	80.9
60	76.4	81.4	86.7
75	80.4	85.6	90.9
90	84.6	88.8	93.6

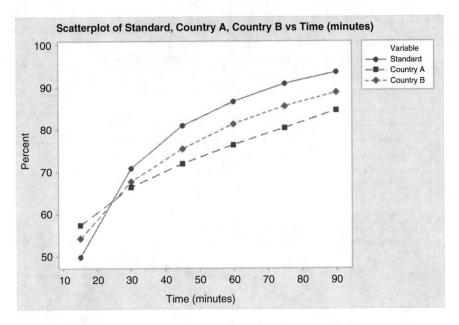

Fig. 7.14 Dissolution profiles comparing two sites to the reference standard

References

Benet LZ, Goyan JE (1995) Bioequivalence and narrow therapeutic index drugs. Pharmacotherapy 15:433–440

FDA (1997) Guidance for industry: dissolution testing of immediate release solid oral dosage forms. In: (BP1), Center for Drug Evaluation and Research. Food and Drug Administration, Rockville, MD, p 9

Federal Register (1995) 60: 61638–61643

Malinowski HJ (2000) Bioavailability and bioequivalency testing. In: Gennaro AR (ed) Chapter 53 in Remington: the science and practice of pharmacy, 20th edn. Lippincott, Williams and Wilkins, Baltimore, p 995

Metzler CM (1974) Bioavailability: a problem in equivalence. Biometrics 30:309–317

Schuirmann DJ (1987) Comparison of the two one-sided tests procedure and the power approach for assessing the equivalence of average bioavailability. J Pharmacokinet Biopharm 15:660

Shah VP et al (1998) In Vitro dissolution profile comparison: statistics and analysis of the similarity factor, f_2. Pharm Res 15:891–898

Chapter 8
Tests to Evaluate Potential Outliers

Abstract This last chapter deals with the evaluation of potential outliers. Outliers are different from out-of-specification data points and represent very extreme aberrant values. Outlies can occur with univariate or multivariate samples and may have disastrous effects on the inferential statistic when analyzing small data sets. Outlier tests are very valuable for in these situations, but may not be needed for larger data sets because the other values will soften the effects of potential outlies on the results of any statistical evaluation. The primary tests described in this chapter are Grubbs' test and Dixon's Q test for univariate situations and an evaluation of the studentized residuals for linear regression-type situations. Minitab applications for evaluating data points as potential outliers are presented.

Keywords Dixon's Q test · Grubbs' test · Hempel's rule · Outliers · Studentized residuals

An outlier, often referred to as an *aberrant value,* is an extreme data point that is significantly different from the remaining values in a set of observations. Based on information, either investigational or statistical, an outlier value may be removed from the data set before performing an inferential test. However, removal of an outlier is discouraged unless the data point can be clearly demonstrated to be erroneous. Outliers can dramatically affect the outcome of a statistical analysis. This is especially true if the sample size is small.

8.1 Regulatory Considerations

Prior to the introduction of USP Chapter <1010> in 2005, there were no compendia guidelines on the treatment of outliers, except for a reference in the beginning of the USP chapter on biological assays (USP 2004 <111>). This lack of guidance or "silence" on the part of USP was noted in the 1993 case of United States versus Barr

© American Association of Pharmaceutical Scientists 2019
J. E. De Muth, *Practical Statistics for Pharmaceutical Analysis*, AAPS
Advances in the Pharmaceutical Sciences Series 40,
https://doi.org/10.1007/978-3-030-33989-0_8

Laboratories, Inc. (USA vs. Barr 1993). Judge Wolin's ruling in the Barr case pointed out the need for compendia guidance in this area of outliers, as well as other analytical measures. USP <1010> attempts to address many of these issues (USP 2011 <1010>).

In 1998 the FDA focused attention on a similar problem, out-of-specification (OOS) test results, and issued a draft guidance (FDA 1998). In addition to outlier tests, the guidance attempts to address retesting, resampling, and averaging of test results. This guidance was updated in 2006, but it ignored <1010>, and no changes were made in the outlier section (FDA 2006).

Similar to other laboratory results, potential outliers must be documented and interpreted. Both USP <1010> and the FDA guidance propose a two-phase approach to identifying and dealing with outliers. When an outlier is suspected, the first phase is a thorough and systematic laboratory investigation to determine if there is a possible assignable cause for the aberrant result. Potential assignable causes include "human error, instrumentation error, calculation error, and product or component deficiency" (USP 2011). If one can identify an assignable cause in the first phase, then the outlier can be removed, and statistically reevaluated with the same sample $(n - 1)$ or the addition of a new replacement sample from the same population is permissible. However, if no assignable cause can be identified, then the second phase is to evaluate the potential aberrant value using statistical outlier tests as part of the overall investigation. When used correctly, the outlier tests described below are valuable statistical tools; however, any judgment about the acceptability of data in which outliers are observed requires careful interpretation.

Determining the most appropriate outlier test will depend on the assumed population distribution and the sample size. If, as the result of either thorough investigation or outlier test, a value is removed as an outlier, the step is termed an "outlier rejection." Both the FDA and USP note that using an outlier test cannot be the sole means for outlier rejection. Even though the outlier tests can be useful as part of the determination of the aberrant nature of a data point, the outlier test can never replace the value of a thorough laboratory investigation. All data, especially outliers, should be kept for future reference. Outliers are not used to calculate the final reportable values but should be footnoted in tables or reports.

An outlier should not be confused with *out-of-specification* (OOS) result. An OOS is a result that falls outside an allowable acceptance criteria that have been established in official compendia and/or company specifications. Outliers are more extreme and more influential than OOS data.

8.2   Univariate Outliers

The most commonly used tests for univariate outliers are Grubbs' test and Dixon's Q-test which are described below and available on Minitab. Both Grubbs' test and Dixon's Q test assume that the population from which the sample is taken is normally distributed. Obviously if data is positively skewed, very large data point would not necessarily be an outlier, only a far point on the positively skewed curve. Either procedure involves testing the most extreme value. If there are potential additional outliers, the second most extreme is tested with $n - 1$ observations (removing the initial outlier).

Other less common tests include a simple huge rule and Hempel's Test.

The larger the sample size, the less the concern about a potential outlier. If doing dissolution testing with only six vessels, a single outlier can have a huge impact on the test results. However, if a sample had 30 or more data points, the effect of a potential outlier would be lessened by the other 29 observations. The author has a simple rule for dealing with potential outliers. Perform the intended statistical analysis both with and without the potential outlier(s), which can be easily accomplished with computer software. If the results of the analysis are the same (rejecting or failing to reject the null hypothesis), the question of whether a value is an outlier becomes a moot issue. In most cases, results for 30 or more observations will not be significantly affected by a single outlier.

8.2.1   The Huge Rule

One method for detecting an aberrant value is to compare the potential outlier to the sample mean and standard deviation with the potential outlier removed from the calculations. This general rule of thumb is to consider the data point as an outlier if that point is located more than four standard deviations from the mean as calculated without the suspected outlier (Marascuilo 1971). The rationale for this rule is that it is extremely unlikely ($p < 0.00005$) to find values more than four standard deviations from the expected center of a normal distribution. The distance, in standard deviations, is measured between the mean and the potential outlier (Fig. 8.1). The calculation uses the sample mean and standard deviation without including the potential outlier. If the resulting value is greater than four (standard deviations from

Huge rule (mean and standard deviation without the potential outlier):

$$M = \frac{|x_i - \bar{X}|}{S}$$

Grubbs' test (smallest value a potential outlier):

$$T = \frac{\bar{X} - x_1}{S}$$

Grubb's test (largest value a potential outlier):

$$T = \frac{x_n - \bar{X}}{S}$$

Fig. 8.1 Calculations for the huge rule and Grubbs' test

the mean), then the data point is considered to be an outlier. Unfortunately Minitab does not include the huge rule, but it can be easily calculated from the descriptive statistics that Minitab can provide.

8.2.2   Grubbs' Test

Grubbs' test involves ranking the observations from smallest to largest $(x_1 < x_2 < x_3 < \dots x_n)$ and calculating the mean and standard deviation for all of the observations in the data set including the potential outlier (Grubbs 1969). This test is also referred to as *extreme studentized deviate test* or *ESD test*. One of the two formulas in Fig. 8.1 is used, depending upon whether x_1 (the smallest value) or x_n (the largest value) is suspected of being a possible outlier. These formulas are occasionally referred to as the *T procedure* or *T method*. The resultant t-value is compared to a critical value on a table of critical values (Grubbs and Beck 1972). However, Minitab reports not only the t-value and corresponding p-value, so a table is not required to evaluate a potential outlier. If the resultant p-value is less than 0.05, the value can be rejected as an outlier. If a second potential outlier is suspected, Grubbs' tests would be performed a second time but without the previous outlier included in the calculation.

8.2.3   Dixon's Q-test

A third method to determine if a suspected value is an outlier in a univariate analysis is to measure the difference between that data point with the next closest value and compare that difference to the total range of observations (Dixon 1953). Various ratios of this type (absolute ratios without regard to sign) make up the Dixon test for outlying observations. The advantage of this test is that the estimation of the standard deviation is not required. Similar to Grubbs' test, the observations are ranked and labeled $(x_1 < x_2 < x_3 < \dots x_{n-2} < x_{n-1} < x_n)$. Formulas for Dixon's test use ratios of ranges and subranges within the data dependent on the sample sizes (Fig. 8.2). The results from the ratio are compared to critical values on a table (Dixon and Massey 1983) and the value rejected as an outlier if it exceeds the critical value. Once again, Minitab provides the q-statistic and corresponding p-value; so a table of critical values is not required. Like Grubbs' test if the p-value is less than 0.05, the observation can be rejected as a statistical outlier. If a second potential outlier is suspected, Dixon's tests would be performed a second time, but without the previous outlier included in the calculation.

Grubbs' and Dixon's tests may not always agree regarding the rejection of the possible outlier, especially when the test statistic results are very close to the allowable error (e.g., 0.05 level). The former involves the mean and standard deviation and the latter a ratio of ranges. Grubbs' test requires more calculations but is considered to be the more powerful of the two tests and can be used when there is more than one suspected outlier (Mason, p. 512).

Fig. 8.2 Various ratios for the Dixon's Q test

Sample Size	Ratio	If x_1 is suspected
$3 \le n \le 7$	τ_{10}	$\dfrac{x_2 - x_1}{x_n - x_1}$
$8 \le n \le 10$	τ_{11}	$\dfrac{x_2 - x_1}{x_{n-1} - x_1}$
$11 \le n \le 13$	τ_{21}	$\dfrac{x_3 - x_1}{x_{n-1} - x_1}$
$14 \le n \le 25$	τ_{22}	$\dfrac{x_3 - x_1}{x_{n-2} - x_1}$

Sample Size	Ratio	If x_n is suspected
$3 \le n \le 7$	τ_{10}	$\dfrac{x_2 - x_{n-1}}{x_n - x_1}$
$8 \le n \le 10$	τ_{11}	$\dfrac{x_n - x_{n-1}}{x_n - x_2}$
$11 \le n \le 13$	τ_{21}	$\dfrac{x_n - x_{n-2}}{x_n - x_2}$
$14 \le n \le 25$	τ_{22}	$\dfrac{x_n - x_{n-2}}{x_n - x_3}$

8.2.4   Hempel's Rule

The underlying assumption with both Grubbs' and Dixon's tests is that the sample being evaluated comes from population with a normal distribution. Hempel's rule for testing outliers is based on the median and can be used for samples from populations with any type of distribution and is not restricted to only normally distributed populations. The test involves calculating the median for the absolute deviation (MAD) value and using this value as the denominator for a ration with the range from the median to the potential outlier in the denominator. This is compared to critical values (Hampel 1985).

8.2.5   Minitab Applications

Minitab provides both Grubbs' test and Dixon's tests, but unfortunately the huge rule and Hempel's rule are not available.

8.2.5.1   Grubbs' Test

Procedure	Stats → Basic Statistics → Outlier test → Options (Grubbs)
Data input	Two options are available: (1) Select the column with the potential outlier "Variable," or (2) select a subset of data with in a column for a multilevel independent variable "By variable."
Options	Automatic default for a 95% confidence interval for a two-tailed test. These can be changed to a different degrees of type I error if required or a one-tailed test is desired.

Graphs   By default a dot plot will be displayed.

Results   By default method, test results and outlier report are provided. Any of these can be removed.

Report   A table reports the results with the mean, standard deviation, the smallest value (min), the largest value (max), the Grubbs statistic, and associated p-value. If "By variable," the results will be reported for each level of the independent variable.

Interpretation An outlier will be present if the p-values is less than 0.05, and text will identify whether the smallest or largest value is the outlier.

8.2.5.2   Dixon's Tests

Procedure   Stats → Basic Statistics → Outlier test → Options (one of Dixon's tests) Six options are available based on the sample size – see Table 8.1.

Data input   Two options are available: (1) Select the column with the potential outlier "Variable" or (2) select a subset of data with in a column for a multilevel independent variable "By variable."

Options   Automatic default for a 95% confidence interval for a two-tailed test. These can be changed to a different degree of type I error if required or a one-tailed test is desired.

Graphs   By default a dot plot will be displayed.

Results   By default method, test results and outlier report are provided. Any of these can be removed.

Report   A table reports the results with the mean, standard deviation, the smallest value (min), the largest value (max), Dixon's statistic, and associated p-value. If "By variable," the results will be reported for each level of the independent variable.

Interpretation An outlier will be present with the p-values is less than 0.05, and text will identify whether the smallest or largest value is the outlier.

Table 8.1 Ranges of sample sizes for various Dixon's ratios

Sample sizes	Dixon's ratio
$3 \leq n \leq 7$	Dixon's Q ratio (Dixon's r_{10})
$8 \leq n \leq 10$	Dixon's r_{11}
$11 \leq n \leq 13$	Dixon's r_{21}
$n \geq 14$	Dixon's r_{22}

8.2.6   Examples

A new automated method is used determining the percent label claim for a batch of compressed tablets. The initial results were (in rank order) 97.98, 99.02, 99.35, 99.45, 99.87, 100.01, 100.11, 100.12, 100.23, 100.25, 100.26, and 100.43%, with a sample mean of 99.757% and standard deviation of 0.705%. Could the smallest value (97.98) be a potential outlier? Minitab output for Grubbs' test is presented in Fig. 8.3. The test statistic is $G = 2.52$, and the associated p-value is 0.025. Therefore, 97.98 can be eliminated as an outlier with greater than 95% confidence in the decision. Similar results are seen with Dixon's r_{21} option in Fig. 8.3 ($r_{21} = 0.60$; $p = 0.043$). Differences in the p-values are due to the fact Grubbs' involves the mean and standard deviation, whereas Dixon's is based on a ratio of ranges.

   With 97.98 eliminated, could the second most extreme value (99.02) also be an outlier? In this case (with $n = 11$), the result would be $G = 2.00$ and $p = 0.290$ for Grubbs' and $r_{21} = 0.35$ and $p = 0.796$ for Dixon's test. In both cases 99.02 cannot be rejected as an outlier.

   The huge rule would produce similar results with the first potential data point being rejected as an outlier ($M = 4.31$, more than four standard deviations from the mean) where the sample mean and standard deviation without the potential outlier are 99.918 ± 0.450. Like the previous tests, the huge rule would fail to identify the second value as an outlier ($M = 2.79$).

   As a default, Minitab also provides a dot plot with the statistical outlier represented by a square point (Fig. 8.4). Visually 97.98 appears distant to the left of the next closest value. This default can be turned off at "Graphs" on the "Outlier Test" panel. Graphically an outlier also could be potentially identified using a box-and-whisker plot with Minitab (Sect. 2.3.2) with outlier(s) symbolized by an asterisk(s). An example for the previous data set, with all 12 observations the box-and-whisker

Grubbs' Test:

Variable	N	Mean	StDev	Min	Max	G	P
%LC	12	99.757	0.705	97.980	100.430	2.52	0.025

Variable	Row	Outlier
%LC	7	97.98

Dixon's Test:

Variable	N	Min	x[2]	x[3]	x[N-2]	x[N-1]	Max	r21	P
%LC	12	97.980	99.020	99.350	100.250	100.260	100.430	0.60	0.043

x[i] denotes the ith smallest observation.

Variable	Row	Outlier
%LC	7	97.98

Fig. 8.3 Minitab output for outlier tests for first univariate example

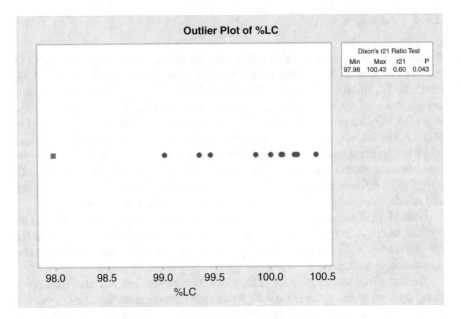

Fig. 8.4 Minitab graphic for the first univariate example

plot (Fig. 8.5), 97.98 appears as an outlier with the asterisk symbol. However, be cautious because these plots tend to be overly generous with assigning outliers symbols, and some noted asterisks will not test positive as outliers with either Grubbs' or Dixon's tests. Graphics should never be used alone to detect outliers. To illustrate this problem with the box-and-whisker plot, consider the data in Table 8.2. If data is plotted, the smallest observation appears to be an outlier in the box-and-whisker plot (Fig. 8.6). However, doing Grubbs' test there is no outlier ($G = 2.73, p = 0.068$), nor does Dixon's test identify an outlier ($r_{22} = 0.39, p = 0.146$). In addition, the huge rule is less than four standard deviations ($M = 3.43$).

8.3   Multivariate Outliers

In the case of correlation or linear regression, where each data point represents values on different axes, an outlier is a point clearly outside the range of the other data points on the respective axis. Outliers may greatly affect the results of the correlation or regression models, especially if there are only a small number of data points. In linear regression-type models, outliers can potentially occur on the dependent variable. A potential outlier in linear regression would be the data point that lies a great distance from the regression line. It can be defined as an observation with an extremely large residual. In contrast, with a correlation model, both variables are dependent and outliers may occur in either variable.

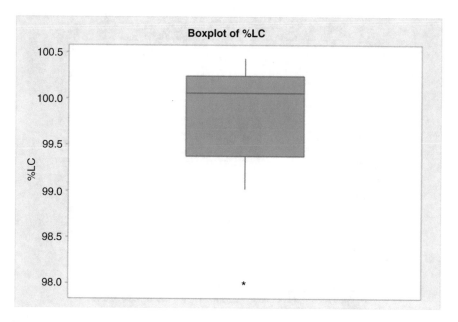

Fig. 8.5 Minitab graphic of a box-and-whisker plot of a potential outlier for first univariate example

Table 8.2 Rank order of data for second univariate example

23.0	25.0
23.8	25.0
23.9	25.1
24.0	25.1
24.1	25.2
24.4	25.2
24.5	25.4
24.8	25.6
24.9	25.9
24.9	26.1
25.0	26.3
25.0	26.6

8.3.1   Linear Regression Outliers

In a regression model where we can control the independent variable and are interested in possible outliers in the dependent (response) variable. Outlier detecting techniques are based on an evaluation of the residuals (distance between the data point and it corresponding point vertically on the regression line).

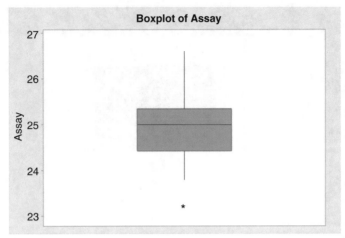

Grubb's test:

Variable	N	Mean	StDev	Min	Max	G	P
Assay	24	24.950	0.833	23.000	26.600	2.34	0.315

Dixon's test:

Variable	N	Min	x[2]	x[3]	x[N-2]	x[N-1]	Max	r22	P
Assay	24	23.000	23.800	23.900	26.100	26.300	26.600	0.29	0.483

Fig. 8.6 Minitab output and box-and-whisker plot and statistical results for second outlier example

Outliers are identified by calculating the corresponding *studentized residuals* and comparing these values to a critical value on *t*-table (Appendix B, Table B2). With this *studentized deleted residual method*, residuals exceeding the critical value would be considered outliers. The formula for calculating the studentized residuals is the distance of each data point from the regression line divided by the square root of the mean square error from the ANOVA table for the regression analysis:

$$t = \frac{y_i - y_c}{\sqrt{MS_E}}$$

As the numerator gets larger (bigger differences) the *t*-value will increase and become more likely to be significant. As seen below Minitab can easily calculate all the studentized residuals; one only needs to determine the critical *t*-value and decide if any of the residuals exceed the critical value. If the critical value is exceeded, the data point can be considered an outlier. A detailed explanation of the studentized deleted residual method is found in Mason et al. (1989) (pp. 518–521).

8.3.2   Correlation Outlier

With correlation an outlier is a data point that falls outside the range of the other points on a scatter plot. Since both the x-axis and y-axis represent dependent variables, an outlier could occur on either axis. A problem occurs when one data point distorts the correlation coefficient. An outlier may have positive or negative effects on the correlation coefficient. If it falls consistently with the other points, it could increase the r-value (Scatterplot A in Fig. 8.7). If it does not follow the trend of the other points, it could greatly decrease the correlation coefficient (Scatterplot B in Fig. 8.7). The smaller the sample size, the greater the effect of the outlier.

There are no set rules for rejecting an outlier with correlation. The best way to spot a potential outlier is to visually inspect a scatter diagram to determine if the data point does not fit the general trend of the other data. One check for a potential outlier is to remove the potential outlier and recalculate the correlation coefficient and determine the influence of the potential outlier on the outcome of the sample. If there are extreme differences between the correlation coefficient with and without the outlier, the data point can be considered an outlier. See the example below.

The decision whether to reject a data point as an outlier is up to the researcher; but he or she must justify deleting any data point to the reader of a technical report or published article.

8.3.3   Minitab Applications

For regression, the first step would be to determine the critical t-value by referring to Table B2 (Appendix B) for $n - 1$ degrees of freedom and desired p-value. Next, identify the studentized residual for each data point by running a linear regression

Stats → Regression → Fit Line Plot

and under the optional buttons select "Storage" and check "Standardized residuals." These residuals are then listed on the Minitab worksheet in the next available column, and residuals are presented on the same rows as their corresponding y-values.

For correlation, view a scatterplot for the data and visually evaluated any potential outlier.

Graph → Scatterplot → Simple

Calculate the correlation coefficient with and without the potential outlier

Stats → Basic Statistics → Correlation

and determine how much the potential outlier effects the correlation coefficient.

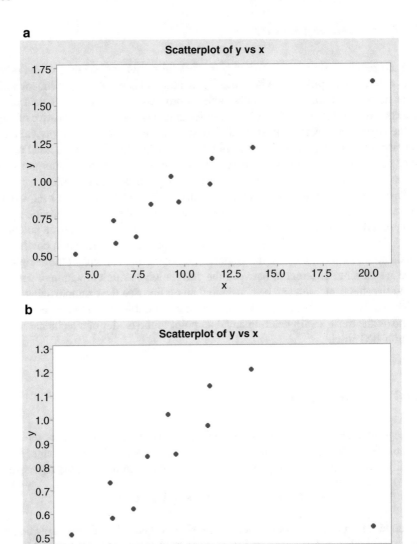

Fig. 8.7 Examples of two correlation distributions

8.3.4   *Examples*

For linear regression consider the following example. During one step in the synthesis of a biological product, there is a brief fermentation period. The concentration (in percent) of one component is evaluated to determine if changes in its concentration will have a linear influence on the yield in units produced. The results of the experiment are presented in Table 8.3. The critical value from Table B2 for 6 ($n - 1$)

Table 8.3 Sample data for
regression outlier testing

Concentration (%)	Units
3.0	99.2
3.5	98.7
4.0	100.2
4.5	113.7
5.0	110.3
5.5	109.4
6.0	115.4

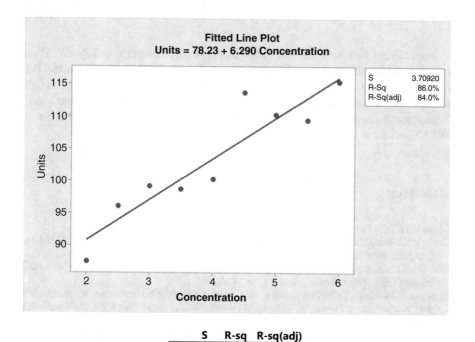

	S	R-sq	R-sq(adj)
	3.70920	86.04%	84.04%

Source	DF	SS	MS	F	P
Regression	1	593.462	593.462	43.14	0.000
Error	7	96.307	13.758		
Total	8	689.769			

Fig. 8.8 Minitab output for linear regression and graphic for best-fit line for yield versus various
concentrations

degrees of freedom and $\alpha = 0.05$ is 2.447. The MS_E used to calculate the studentized
residuals is 13.758 from the ANOVA table for linear regression, and the potential
outlier is at concentration 4.5% (Fig. 8.8). However, looking at the outcomes from
Minitab, storing the studentized residuals back on the worksheet (Fig. 8.9), none of
the values exceed 2.447. This includes 4.5%, thus no outliers.

Fig. 8.9 Minitab storage
of studentized residuals
(*C*3) for the various
concentrations

C1	C2	C3
Concentration	Units	SRES1
3.0	99.2	0.62470
3.5	98.7	-0.44568
4.0	100.2	-0.91187
4.5	113.7	2.06866
5.0	110.3	0.18467
5.5	109.4	-1.07386
6.0	115.4	-0.19443

For correlation, reconsider two scenarios previously illustrated in Fig. 8.7. There is a positive correlation for the 10 data points left side of both scatterplot A and B is $r = +0.935, p < 0.001$. Considering all 11 points in Scatterplot A, there is only a slight increase in the correlation coefficient to $r = +0.969$, $p < 0.001$. However, with Scatterplot B the potential outlier substantially drops the correlation coefficient to $+0.247$, and the $p = 0.464$ would most likely be deemed an outlier.

References

Dixon WJ (1953) Processing data for outliers. Biometrics 1:74–89

Dixon WJ, Massey FJ (1983) Introduction to statistical analysis (table A-8e). McGraw-Hill Book Company, New York

FDA Draft Guidance (1998) Investigating out of specification (OOS) test results for pharmaceutical production, guidance for industry. FDA, Rockville

FDA Draft Guidance (2006) Investigating out of specification (OOS) test results for pharmaceutical production, guidance for industry. FDA, Rockville. (https://www.fda.gov/downloads/Drugs/Guidances/ucm070287.pdf)

Grubbs FE (1969) Procedures for detecting outlying observations in samples. Technometrics 11:1–21

Grubbs FE, Beck G (1972) Extension of sample size and percentage points for significance tests of outlying observations. Technometrics 14:847–854

Hampel FR (1985) The breakdown points of the mean combined with some rejection rules. Technometrics 27:95–107

Marascuilo LA (1971) Statistical methods for behavioral science research. McGraw Hill, New York, p 199

Mason RL, Gunst RF, Hess JL (1989) Statistical design and analysis of experiments. John Wiley and Sons, New York

United States v. Barr Laboratories, Inc., 812 F. Supp. 458, 1993 (U.S. D.C. New Jersey)

USP (2004). <111> Design and analysis of biological assays, United States Pharmacopeia/National Formulary, Rockville, pp. 2180–2221

USP (2011). <1010> Analytical data—interpretation and treatment, United States Pharmacopeia/National Formulary, Rockville, pp. 419–430

Appendices

Appendix A: Flow Charts for the Selection of the Most Appropriate Inferential Test Given the Types of Variables in the Study

For any given hypothesis being tested, the researcher must first identify the independent variable(s) and/or dependent variable(s). This begins the process seen in Panel A. Next the researcher must determine if the data presented by the respective variables involves discrete or continuous data (D/C?). Lastly, at end points in the decision-making process, the researcher must determine if the sample data comes from populations that are normally distributed and if there is more than one level of a discrete independent variable does there appear to be homogeneity of variance (ND/H?).

© American Association of Pharmaceutical Scientists 2019
J. E. De Muth, *Practical Statistics for Pharmaceutical Analysis*, AAPS
Advances in the Pharmaceutical Sciences Series 40,
https://doi.org/10.1007/978-3-030-33989-0

Panel A

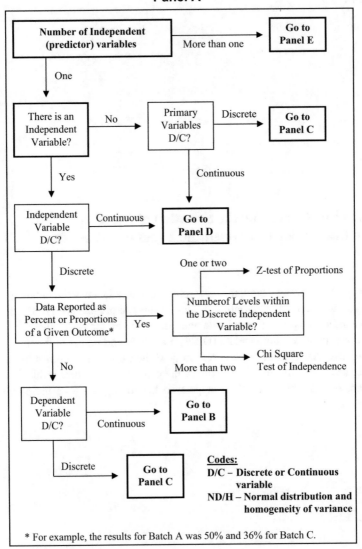

* For example, the results for Batch A was 50% and 36% for Batch C.

Panel B

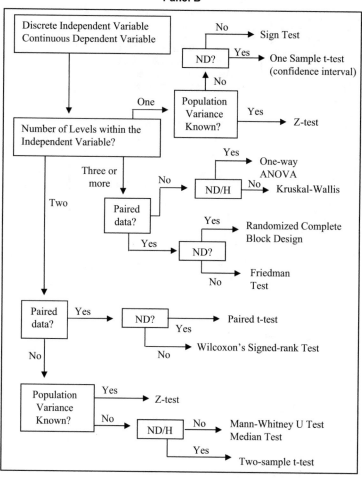

Panel C

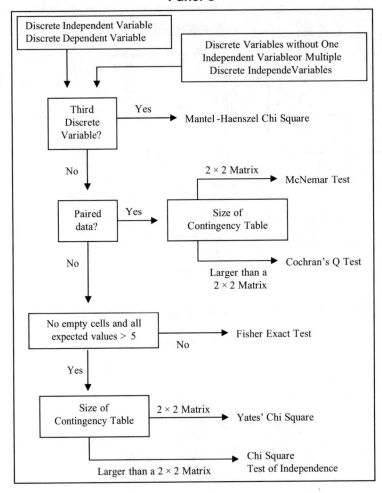

Panel D

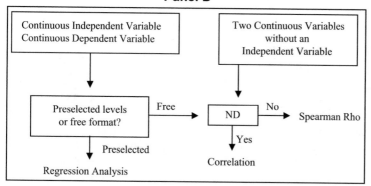

Panel E

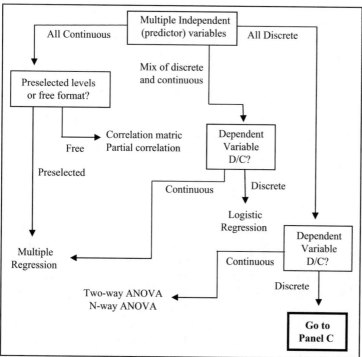

Appendix B: Common Statistical Tables

Table B1 Normal standardized distribution

(Area under the curve between 0 and z)

z	0.00	0.01	0.02	0.03	0.04	0.05	0.06	0.07	0.08	0.09
0.0	0.0000	0.0040	0.0080	0.0120	0.0160	0.0199	0.0239	0.0279	0.0319	0.0359
0.1	0.0398	0.0438	0.0478	0.0517	0.0557	0.0596	0.0636	0.0675	0.0714	0.0753
0.2	0.0793	0.0832	0.0871	0.0910	0.0948	0.0987	0.1026	0.1064	0.1103	0.1141
0.3	0.1179	0.1217	0.1255	0.1293	0.1331	0.1368	0.1406	0.1443	0.1480	0.1517
0.4	0.1554	0.1591	0.1628	0.1664	0.1700	0.1736	0.1772	0.1808	0.1844	0.1879
0.5	0.1915	0.1950	0.1985	0.2019	0.2054	0.2088	0.2123	0.2157	0.2190	0.2224
0.6	0.2257	0.2291	0.2324	0.2357	0.2389	0.2422	0.2454	0.2486	0.2517	0.2549
0.7	0.2580	0.2611	0.2642	0.2673	0.2704	0.2734	0.2764	0.2794	0.2823	0.2852
0.8	0.2881	0.2910	0.2939	0.2967	0.2995	0.3023	0.3051	0.3078	0.3106	0.3133
0.9	0.3159	0.3186	0.3212	0.3238	0.3264	0.3289	0.3315	0.3340	0.3365	0.3389
1.0	0.3413	0.3438	0.3461	0.3485	0.3508	0.3531	0.3554	0.3577	0.3599	0.3621
1.1	0.3643	0.3665	0.3686	0.3708	0.3729	0.3749	0.3770	0.3790	0.3810	0.3830
1.2	0.3849	0.3869	0.3888	0.3907	0.3925	0.3944	0.3962	0.3980	0.3997	0.4015
1.3	0.4032	0.4049	0.4066	0.4082	0.4099	0.4115	0.4131	0.4147	0.4162	0.4177
1.4	0.4192	0.4207	0.4222	0.4236	0.4251	0.4265	0.4279	0.4292	0.4306	0.4319
1.5	0.4332	0.4345	0.4357	0.4370	0.4382	0.4394	0.4406	0.4418	0.4429	0.4441
1.6	0.4452	0.4463	0.4474	0.4484	0.4495	0.4505	0.4515	0.4525	0.4535	0.4545
1.7	0.4554	0.4564	0.4573	0.4582	0.4591	0.4599	0.4608	0.4616	0.4625	0.4633
1.8	0.4641	0.4649	0.4656	0.4664	0.4671	0.4678	0.4686	0.4693	0.4699	0.4706
1.9	0.4713	0.4719	0.4726	0.4732	0.4738	0.4744	0.4750	0.4756	0.4761	0.4767
2.0	0.4772	0.4778	0.4783	0.4788	0.4793	0.4798	0.4803	0.4808	0.4812	0.4817
2.1	0.4821	0.4826	0.4830	0.4834	0.4838	0.4842	0.4846	0.4850	0.4854	0.4857
2.2	0.4861	0.4864	0.4868	0.4871	0.4875	0.4878	0.4881	0.4884	0.4887	0.4890
2.3	0.4893	0.4896	0.4898	0.4901	0.4904	0.4906	0.4909	0.4911	0.4913	0.4916
2.4	0.4918	0.4920	0.4922	0.4925	0.4927	0.4929	0.4931	0.4932	0.4934	0.4936
2.5	0.4938	0.4940	0.4941	0.4943	0.4945	0.4946	0.4948	0.4949	0.4951	0.4952
2.6	0.4953	0.4955	0.4956	0.4957	0.4959	0.4960	0.4961	0.4962	0.4963	0.4964
2.7	0.4965	0.4966	0.4967	0.4968	0.4969	0.4970	0.4971	0.4972	0.4973	0.4974
2.8	0.4974	0.4975	0.4976	0.4977	0.4977	0.4978	0.4979	0.4979	0.4980	0.4981
2.9	0.4981	0.4982	0.4982	0.4983	0.4984	0.4984	0.4985	0.4985	0.4986	0.4986
3.0	0.4987	0.4987	0.4987	0.4988	0.4988	0.4989	0.4989	0.4989	0.4990	0.4990
3.1	0.4990	0.4991	0.4991	0.4991	0.4992	0.4992	0.4992	0.4992	0.4993	0.4993
3.2	0.4993	0.4993	0.4994	0.4994	0.4994	0.4994	0.4994	0.4995	0.4995	0.4995
3.3	0.4995	0.4995	0.4995	0.4996	0.4996	0.4996	0.4996	0.4996	0.4996	0.4997
3.4	0.4997	0.4997	0.4997	0.4997	0.4997	0.4997	0.4997	0.4997	0.4997	0.4998
3.5	0.4998	0.4998	0.4998	0.4998	0.4998	0.4998	0.4998	0.4998	0.4998	0.4998
3.6	0.4998	0.4998	0.4999	0.4999	0.4999	0.4999	0.4999	0.4999	0.4999	0.4999

This table was created with Microsoft® Excel 2010 using function command NORM.S.DIST(*value*)-0.5

Table B2 Student t-distribution $(1 - \alpha/2)$

d.f.	$t_{0.80}$	$t_{0.90}$	$t_{0.95}$	$t_{0.975}$	$t_{0.99}$	$t_{0.995}$	$t_{0.9975}$	$t_{0.9995}$
1	0.7265	3.0777	6.3137	12.706	31.821	63.656	127.32	636.58
2	0.6172	1.8856	2.9200	4.3027	6.9645	9.9250	14.089	31.600
3	0.5844	1.6377	2.3534	3.1824	4.5407	5.8408	7.4532	12.924
4	0.5686	1.5332	2.1318	2.7765	3.7469	4.6041	5.5975	8.6101
5	0.5594	1.4759	2.0150	2.5706	3.3649	4.0321	4.7733	6.8685
6	0.5534	1.4398	1.9432	2.4469	3.1427	3.7074	4.3168	5.9587
7	0.5491	1.4149	1.8946	2.3646	2.9979	3.4995	4.0294	5.4081
8	0.5459	1.3968	1.8595	2.3060	2.8965	3.3554	3.8325	5.0414
9	0.5435	1.3830	1.8331	2.2622	2.8214	3.2498	3.6896	4.7809
10	0.5415	1.3722	1.8125	2.2281	2.7638	3.1693	3.5814	4.5868
11	0.5399	1.3634	1.7959	2.2010	2.7181	3.1058	3.4966	4.4369
12	0.5386	1.3562	1.7823	2.1788	2.6810	3.0545	3.4284	4.3178
13	0.5375	1.3502	1.7709	2.1604	2.6503	3.0123	3.3725	4.2209
14	0.5366	1.3450	1.7613	2.1448	2.6245	2.9768	3.3257	4.1403
15	0.5357	1.3406	1.7531	2.1315	2.6025	2.9467	3.2860	4.0728
16	0.5350	1.3368	1.7459	2.1199	2.5835	2.9208	3.2520	4.0149
17	0.5344	1.3334	1.7396	2.1098	2.5669	2.8982	3.2224	3.9651
18	0.5338	1.3304	1.7341	2.1009	2.5524	2.8784	3.1966	3.9217
19	0.5333	1.3277	1.7291	2.0930	2.5395	2.8609	3.1737	3.8833
20	0.5329	1.3253	1.7247	2.0860	2.5280	2.8453	3.1534	3.8496
21	0.5325	1.3232	1.7207	2.0796	2.5176	2.8314	3.1352	3.8193
22	0.5321	1.3212	1.7171	2.0739	2.5083	2.8188	3.1188	3.7922
23	0.5317	1.3195	1.7139	2.0687	2.4999	2.8073	3.1040	3.7676
24	0.5314	1.3178	1.7109	2.0639	2.4922	2.7970	3.0905	3.7454
25	0.5312	1.3163	1.7081	2.0595	2.4851	2.7874	3.0782	3.7251
30	0.5300	1.3104	1.6973	2.0423	2.4573	2.7500	3.0298	3.6460
40	0.5286	1.3031	1.6839	2.0211	2.4233	2.7045	2.9712	3.5510
50	0.5278	1.2987	1.6759	2.0086	2.4033	2.6778	2.9370	3.4960
60	0.5272	1.2958	1.6706	2.0003	2.3901	2.6603	2.9146	3.4602
80	0.5265	1.2922	1.6641	1.9901	2.3739	2.6387	2.8870	3.4164
100	0.5261	1.2901	1.6602	1.9840	2.3642	2.6259	2.8707	3.3905
120	0.5258	1.2886	1.6576	1.9799	2.3578	2.6174	2.8599	3.3734
160	0.5254	1.2869	1.6544	1.9749	2.3499	2.6069	2.8465	3.3523
200	0.5252	1.2858	1.6525	1.9719	2.3451	2.6006	2.8385	3.3398
∞	0.5244	1.2816	1.6450	1.9602	2.3267	2.5763	2.8076	3.2915

This table was created with Microsoft® Excel 2010, function command T.INV.2T (alpha,df)

Table B3 Analysis of variance F-distribution

ν_1	ν_2	$F_{0.80}$	$F_{0.90}$	$F_{0.95}$	$F_{0.975}$	$F_{0.99}$	$F_{0.999}$	$F_{0.9999}$
	1	9.4722	39.864	161.45	647.79	4052.2	4×10^5	4×10^7
	2	3.5556	8.5263	18.513	38.506	98.502	998.38	1×10^4
	3	2.6822	5.5383	10.128	17.443	34.116	167.06	784.17
	4	2.3507	4.5448	7.7086	12.218	21.198	74.127	241.68
	5	2.1782	4.0604	6.6079	10.007	16.258	47.177	124.80
	6	2.0729	3.7760	5.9874	8.8131	13.745	35.507	82.422
	7	2.0020	3.5894	5.5915	8.0727	12.246	29.246	62.166
	8	1.9511	3.4579	5.3176	7.5709	11.259	25.415	50.699
	9	1.9128	3.3603	5.1174	7.2093	10.562	22.857	43.481
	10	1.8829	3.2850	4.9646	6.9367	10.044	21.038	38.592
	11	1.8589	3.2252	4.8443	6.7241	9.6461	19.687	35.041
	12	1.8393	3.1766	4.7472	6.5538	9.3303	18.645	32.422
	13	1.8230	3.1362	4.6672	6.4143	9.0738	17.815	30.384
1	14	1.8091	3.1022	4.6001	6.2979	8.8617	17.142	28.755
	15	1.7972	3.0732	4.5431	6.1995	8.6832	16.587	27.445
	16	1.7869	3.0481	4.4940	6.1151	8.5309	16.120	26.368
	17	1.7779	3.0262	4.4513	6.0420	8.3998	15.722	25.437
	18	1.7699	3.0070	4.4139	5.9781	8.2855	15.380	24.651
	19	1.7629	2.9899	4.3808	5.9216	8.1850	15.081	23.982
	20	1.7565	2.9747	4.3513	5.8715	8.0960	14.819	23.399
	22	1.7457	2.9486	4.3009	5.7863	7.9453	14.381	22.439
	24	1.7367	2.9271	4.2597	5.7166	7.8229	14.028	21.653
	26	1.7292	2.9091	4.2252	5.6586	7.7213	13.739	21.042
	30	1.7172	2.8807	4.1709	5.5675	7.5624	13.293	20.096
	35	1.7062	2.8547	4.1213	5.4848	7.4191	12.897	19.267
	40	1.6980	2.8353	4.0847	5.4239	7.3142	12.609	18.670
	45	1.6917	2.8205	4.0566	5.3773	7.2339	12.393	18.219
	50	1.6867	2.8087	4.0343	5.3403	7.1706	12.222	17.884
	60	1.6792	2.7911	4.0012	5.2856	7.0771	11.973	17.375
	90	1.6668	2.7621	3.9469	5.1962	6.9251	11.573	16.589
	120	1.6606	2.7478	3.9201	5.1523	6.8509	11.380	16.204
	240	1.6515	2.7266	3.8805	5.0875	6.7416	11.099	15.658
	∞	1.6423	2.7053	3.8415	5.0239	6.6349	10.828	15.134
	2	4.000	9.000	19.00	39.00	99.00	998.8	1×10^4
	3	2.886	5.462	9.552	16.04	30.82	148.5	694.8
	4	2.472	4.325	6.944	10.65	18.00	61.25	197.9
	5	2.259	3.780	5.786	8.434	13.27	37.12	97.09
	6	2.130	3.463	5.143	7.260	10.92	27.00	61.58
	8	1.981	3.113	4.459	6.059	8.649	18.49	35.97
2	10	1.899	2.924	4.103	5.456	7.559	14.90	26.54
	12	1.846	2.807	3.885	5.096	6.927	12.97	21.86

(continued)

Table B3 (continued)

ν_1	ν_2	$F_{0.80}$	$F_{0.90}$	$F_{0.95}$	$F_{0.975}$	$F_{0.99}$	$F_{0.999}$	$F_{0.9999}$
	15	1.795	2.695	3.682	4.765	6.359	11.34	18.10
	20	1.746	2.589	3.493	4.461	5.849	9.953	15.12
	24	1.722	2.538	3.403	4.319	5.614	9.340	13.85
	30	1.699	2.489	3.316	4.182	5.390	8.773	12.72
	40	1.676	2.440	3.232	4.051	5.178	8.251	11.70
	60	1.653	2.393	3.150	3.925	4.977	7.768	10.78
	120	1.631	2.347	3.072	3.805	4.787	7.321	9.954
	∞	1.609	2.303	2.996	3.689	4.605	6.908	9.211
	2	4.1563	9.1618	19.164	39.166	99.164	999.31	1×10^4
	3	2.9359	5.3908	9.2766	15.439	29.457	141.10	659.38
	4	2.4847	4.1909	6.5914	9.9792	16.694	56.170	181.14
	5	2.2530	3.6195	5.4094	7.7636	12.060	33.200	86.380
	6	2.1126	3.2888	4.7571	6.5988	9.7796	23.705	53.667
	8	1.9513	2.9238	4.0662	5.4160	7.5910	15.829	30.443
3	10	1.8614	2.7277	3.7083	4.8256	6.5523	12.553	22.032
	12	1.8042	2.6055	3.4903	4.4742	5.9525	10.805	17.899
	15	1.7490	2.4898	3.2874	4.1528	5.4170	9.3351	14.639
	20	1.6958	2.3801	3.0984	3.8587	4.9382	8.0981	12.049
	24	1.6699	2.3274	3.0088	3.7211	4.7181	7.5543	10.965
	30	1.6445	2.2761	2.9223	3.5893	4.5097	7.0545	9.9972
	40	1.6195	2.2261	2.8387	3.4633	4.3126	6.5947	9.1277
	60	1.5950	2.1774	2.7581	3.3425	4.1259	6.1714	8.3528
	120	1.5709	2.1300	2.6802	3.2269	3.9491	5.7812	7.6579
	∞	1.5472	2.0838	2.6049	3.1162	3.7816	5.4220	7.0359
	2	4.2361	9.2434	19.247	39.248	99.251	999.31	1×10^4
	3	2.9555	5.3427	9.1172	15.101	28.710	137.08	640.75
	4	2.4826	4.1072	6.3882	9.6045	15.977	53.435	171.83
	5	2.2397	3.5202	5.1922	7.3879	11.392	31.083	80.559
	6	2.0924	3.1808	4.5337	6.2271	9.1484	21.922	49.418
	8	1.9230	2.8064	3.8379	5.0526	7.0061	14.392	27.474
4	10	1.8286	2.6053	3.4780	4.4683	5.9944	11.283	19.631
	12	1.7684	2.4801	3.2592	4.1212	5.4119	9.6334	15.789
	15	1.7103	2.3614	3.0556	3.8043	4.8932	8.2528	12.777
	20	1.6543	2.2489	2.8661	3.5147	4.4307	7.0959	10.419
	24	1.6269	2.1949	2.7763	3.3794	4.2185	6.5893	9.4224
	30	1.6001	2.1422	2.6896	3.2499	4.0179	6.1245	8.5420
	40	1.5737	2.0909	2.6060	3.1261	3.8283	5.6980	7.7598
	60	1.5478	2.0410	2.5252	3.0077	3.6491	5.3069	7.0577
	120	1.5222	1.9923	2.4472	2.8943	3.4795	4.9472	6.4356
	∞	1.4972	1.9449	2.3719	2.7858	3.3192	4.6166	5.8790
	2	4.2844	9.2926	19.296	39.298	99.302	999.31	1×10^4

(continued)

Table B3 (continued)

ν_1	ν_2	$F_{0.80}$	$F_{0.90}$	$F_{0.95}$	$F_{0.975}$	$F_{0.99}$	$F_{0.999}$	$F_{0.9999}$
	3	2.9652	5.3091	9.0134	14.885	28.237	134.58	627.71
	4	2.4780	4.0506	6.2561	9.3645	15.522	51.718	166.24
	5	2.2275	3.4530	5.0503	7.1464	10.967	29.751	76.834
	6	2.0755	3.1075	4.3874	5.9875	8.7459	20.802	46.741
	8	1.9005	2.7264	3.6875	4.8173	6.6318	13.484	25.640
5	10	1.8027	2.5216	3.3258	4.2361	5.6364	10.481	18.132
	12	1.7403	2.3940	3.1059	3.8911	5.0644	8.8921	14.465
	15	1.6801	2.2730	2.9013	3.5764	4.5556	7.5670	11.627
	20	1.6218	2.1582	2.7109	3.2891	4.1027	6.4606	9.3860
	24	1.5933	2.1030	2.6207	3.1548	3.8951	5.9767	8.4547
	30	1.5654	2.0492	2.5336	3.0265	3.6990	5.5338	7.6325
	40	1.5379	1.9968	2.4495	2.9037	3.5138	5.1282	6.8976
	60	1.5108	1.9457	2.3683	2.7863	3.3389	4.7567	6.2464
	120	1.4841	1.8959	2.2899	2.6740	3.1735	4.4156	5.6662
	∞	1.4579	1.8473	2.2141	2.5665	3.0172	4.1030	5.1477
	2	4.3168	9.3255	19.329	39.331	99.331	999.31	1×10^4
	3	2.9707	5.2847	8.9407	14.735	27.911	132.83	620.26
	4	2.4733	4.0097	6.1631	9.1973	15.207	50.524	162.05
	5	2.2174	3.4045	4.9503	6.9777	10.672	28.835	74.506
	6	2.0619	3.0546	4.2839	5.8197	8.4660	20.031	44.936
	8	1.8826	2.6683	3.5806	4.6517	6.3707	12.858	24.360
6	10	1.7823	2.4606	3.2172	4.0721	5.3858	9.9262	17.084
	12	1.7182	2.3310	2.9961	3.7283	4.8205	8.3783	13.562
	15	1.6561	2.2081	2.7905	3.4147	4.3183	7.0913	10.819
	20	1.5960	2.0913	2.5990	3.1283	3.8714	6.0186	8.6802
	24	1.5667	2.0351	2.5082	2.9946	3.6667	5.5506	7.7926
	30	1.5378	1.9803	2.4205	2.8667	3.4735	5.1223	6.9995
	40	1.5093	1.9269	2.3359	2.7444	3.2910	4.7307	6.3010
	60	1.4813	1.8747	2.2541	2.6274	3.1187	4.3719	5.6825
	120	1.4536	1.8238	2.1750	2.5154	2.9559	4.0436	5.1332
	∞	1.4263	1.7741	2.0986	2.4082	2.8020	3.7430	4.6421
	2	4.3401	9.3491	19.353	39.356	99.357	999.31	1×10^4
	3	2.9741	5.2662	8.8867	14.624	27.671	131.61	614.67
	4	2.4691	3.9790	6.0942	9.0741	14.976	49.651	159.26
	5	2.2090	3.3679	4.8759	6.8530	10.456	28.165	72.643
	6	2.0508	3.0145	4.2067	5.6955	8.2600	19.463	43.539
	8	1.8682	2.6241	3.5005	4.5285	6.1776	12.398	23.429
7	10	1.7658	2.4140	3.1355	3.9498	5.2001	9.5170	16.327
	12	1.7003	2.2828	2.9134	3.6065	4.6395	8.0008	12.893
	15	1.6368	2.1582	2.7066	3.2934	4.1416	6.7412	10.230
	20	1.5752	2.0397	2.5140	3.0074	3.6987	5.6921	8.1563

(continued)

Table B3 (continued)

ν_1	ν_2	$F_{0.80}$	$F_{0.90}$	$F_{0.95}$	$F_{0.975}$	$F_{0.99}$	$F_{0.999}$	$F_{0.9999}$
	24	1.5451	1.9826	2.4226	2.8738	3.4959	5.2351	7.2978
	30	1.5154	1.9269	2.3343	2.7460	3.3045	4.8171	6.5374
	40	1.4861	1.8725	2.2490	2.6238	3.1238	4.4356	5.8644
	60	1.4572	1.8194	2.1665	2.5068	2.9530	4.0864	5.2678
	120	1.4287	1.7675	2.0868	2.3948	2.7918	3.7669	4.7385
	∞	1.4005	1.7167	2.0096	2.2875	2.6393	3.4745	4.2673
	5	2.2021	3.3393	4.8183	6.7572	10.289	27.649	71.246
	10	1.7523	2.3771	3.0717	3.8549	5.0567	9.2041	15.745
	15	1.6209	2.1185	2.6408	3.1987	4.0044	6.4706	9.7789
	20	1.5580	1.9985	2.4471	2.9128	3.5644	5.4401	7.7562
8	30	1.4968	1.8841	2.2662	2.6513	3.1726	4.5816	6.1809
	40	1.4668	1.8289	2.1802	2.5289	2.9930	4.2071	5.5261
	60	1.4371	1.7748	2.0970	2.4117	2.8233	3.8649	4.9477
	120	1.4078	1.7220	2.0164	2.2994	2.6629	3.5518	4.4329
	∞	1.3788	1.6702	1.9384	2.1918	2.5113	3.2655	3.9781
	5	2.1963	3.3163	4.7725	6.6810	10.158	27.241	70.082
	10	1.7411	2.3473	3.0204	3.7790	4.9424	8.9558	15.280
	15	1.6076	2.0862	2.5876	3.1227	3.8948	6.2560	9.4224
	20	1.5436	1.9649	2.3928	2.8365	3.4567	5.2391	7.4397
9	30	1.4812	1.8490	2.2107	2.5746	3.0665	4.3929	5.8972
	40	1.4505	1.7929	2.1240	2.4519	2.8876	4.0243	5.2569
	60	1.4201	1.7380	2.0401	2.3344	2.7185	3.6873	4.6912
	120	1.3901	1.6842	1.9588	2.2217	2.5586	3.3792	4.1910
	∞	1.3602	1.6315	1.8799	2.1136	2.4073	3.0975	3.7471
	5	2.1914	3.2974	4.7351	6.6192	10.051	26.914	69.267
	10	1.7316	2.3226	2.9782	3.7168	4.8491	8.7539	14.901
	15	1.5964	2.0593	2.5437	3.0602	3.8049	6.0809	9.1313
	20	1.5313	1.9367	2.3479	2.7737	3.3682	5.0754	7.1814
10	30	1.4678	1.8195	2.1646	2.5112	2.9791	4.2387	5.6643
	40	1.4365	1.7627	2.0773	2.3882	2.8005	3.8744	5.0350
	60	1.4055	1.7070	1.9926	2.2702	2.6318	3.5416	4.4820
	120	1.3748	1.6524	1.9105	2.1570	2.4721	3.2371	3.9909
	∞	1.3442	1.5987	1.8307	2.0483	2.3209	2.9588	3.5561
	5	2.1835	3.2682	4.6777	6.5245	9.8883	26.419	67.987
	10	1.7164	2.2841	2.9130	3.6210	4.7058	8.4456	14.334
	15	1.5782	2.0171	2.4753	2.9633	3.6662	5.8121	8.6875
	20	1.5115	1.8924	2.2776	2.6758	3.2311	4.8231	6.7812
12	30	1.4461	1.7727	2.0921	2.4120	2.8431	4.0006	5.3078
	40	1.4137	1.7146	2.0035	2.2882	2.6648	3.6425	4.6966
	60	1.3816	1.6574	1.9174	2.1692	2.4961	3.3153	4.1582
	120	1.3496	1.6012	1.8337	2.0548	2.3363	3.0161	3.6816

(continued)

Table B3 (continued)

ν_1	ν_2	$F_{0.80}$	$F_{0.90}$	$F_{0.95}$	$F_{0.975}$	$F_{0.99}$	$F_{0.999}$	$F_{0.9999}$
	∞	1.3177	1.5458	1.7522	1.9447	2.1847	2.7425	3.2614
	5	2.1751	3.2380	4.6188	6.4277	9.7223	25.910	66.590
	10	1.7000	2.2435	2.8450	3.5217	4.5582	8.1291	13.752
	15	1.5584	1.9722	2.4034	2.8621	3.5222	5.5352	8.2291
	20	1.4897	1.8449	2.2033	2.5731	3.0880	4.5616	6.3737
15	30	1.4220	1.7223	2.0148	2.3072	2.7002	3.7528	4.9386
	40	1.3883	1.6624	1.9245	2.1819	2.5216	3.4004	4.3456
	60	1.3547	1.6034	1.8364	2.0613	2.3523	3.0782	3.8217
	120	1.3211	1.5450	1.7505	1.9450	2.1915	2.7833	3.3597
	∞	1.2874	1.4871	1.6664	1.8326	2.0385	2.5132	2.9504
	5	2.1660	3.2067	4.5581	6.3285	9.5527	25.393	65.193
	10	1.6823	2.2007	2.7740	3.4185	4.4054	7.8035	13.155
	15	1.5367	1.9243	2.3275	2.7559	3.3719	5.2487	7.7562
	20	1.4656	1.7938	2.1242	2.4645	2.9377	4.2901	5.9517
20	30	1.3949	1.6673	1.9317	2.1952	2.5487	3.4927	4.5547
	40	1.3596	1.6052	1.8389	2.0677	2.3689	3.1450	3.9763
	60	1.3241	1.5435	1.7480	1.9445	2.1978	2.8265	3.4688
	120	1.2882	1.4821	1.6587	1.8249	2.0346	2.5344	3.0177
	∞	1.2519	1.4206	1.5705	1.7085	1.8783	2.2658	2.6193

This table was created using Microsoft® Excel 2010, function command F.INV.RT (alpha, df1, df2)

Table B4 Critical values for r (correlation coefficient)

d.f.	0.01	0.05	0.01	0.001
1	0.988	0.997	0.999	1.00
2	0.900	0.950	0.990	0.999
3	0.805	0.878	0.959	0.991
4	0.730	0.811	0.917	0.974
5	0.669	0.755	0.875	0.951
6	0.622	0.707	0.834	0.925
7	0.582	0.666	0.798	0.898
8	0.549	0.632	0.765	0.872
9	0.521	0.602	0.735	0.847
10	0.497	0.576	0.708	0.823
11	0.476	0.553	0.684	0.801
12	0.458	0.532	0.661	0.780
13	0.441	0.514	0.641	0.760
14	0.426	0.497	0.623	0.742
15	0.412	0.482	0.606	0.725
16	0.400	0.468	0.590	0.708
17	0.389	0.456	0.575	0.693
18	0.378	0.444	0.561	0.679
19	0.369	0.433	0.549	0.665
20	0.360	0.423	0.537	0.652
25	0.323	0.381	0.487	0.597
30	0.296	0.349	0.449	0.554
35	0.275	0.325	0.418	0.519
40	0.257	0.304	0.393	0.490
50	0.231	0.273	0.354	0.443
60	0.211	0.250	0.325	0.408
80	0.183	0.217	0.283	0.357
100	0.164	0.195	0.254	0.321
150	0.134	0.159	0.208	0.264
200	0.116	0.138	0.181	0.230

This table was created using Microsoft® Excel 2010, function command T.INV.2T (alpha,df) to obtain t-critical for formula r-critical $= r_{critical} = \sqrt{\left(t_{critical}\right)^2 / \left[\left(t_{critical}\right)^2 + df\right]}$

Table B5 Chi square distribution

d.f.	$\alpha = 0.10$	0.05	0.025	0.01	0.005	0.001	0.0001
1	2.7055	3.8415	5.0239	6.6349	7.8794	10.827	15.134
2	4.6052	5.9915	7.3778	9.2104	10.597	13.815	18.425
3	6.2514	7.8147	9.3484	11.345	12.838	16.266	21.104
4	7.7794	9.4877	11.143	13.277	14.860	18.466	23.506
5	9.2363	11.070	12.832	15.086	16.750	20.515	25.751
6	10.645	12.592	14.449	16.812	18.548	22.457	27.853
7	12.017	14.067	16.013	18.475	20.278	24.321	29.881
8	13.362	15.507	17.535	20.090	21.955	26.124	31.827
9	14.684	16.919	19.023	21.666	23.589	27.877	33.725
10	15.987	18.307	20.483	23.209	25.188	29.588	35.557
11	17.275	19.675	21.920	24.725	26.757	31.264	37.365
12	18.549	21.026	23.337	26.217	28.300	32.909	39.131
13	19.812	22.362	24.736	27.688	29.819	34.527	40.873
14	21.064	23.685	26.119	29.141	31.319	36.124	42.575
15	22.307	24.996	27.488	30.578	32.801	37.698	44.260
16	23.542	26.296	28.845	32.000	34.267	39.252	45.926
17	24.769	27.587	30.191	33.409	35.718	40.791	47.559
18	25.989	28.869	31.526	34.805	37.156	42.312	49.185
19	27.204	30.144	32.852	36.191	38.582	43.819	50.787
20	28.412	31.410	34.170	37.566	39.997	45.314	52.383
21	29.615	32.671	35.479	38.932	41.401	46.796	53.960
22	30.813	33.924	36.781	40.289	42.796	48.268	55.524
23	32.007	35.172	38.076	41.638	44.181	49.728	57.067
24	33.196	36.415	39.364	42.980	45.558	51.179	58.607
25	34.382	37.652	40.646	44.314	46.928	52.619	60.136

This table was created with Microsoft® Excel 2010, function command CHI.INV.RT (alpha,df)

Appendix C: Summary of Initial Commands for Minitab 19®

ANOVA, one-way design (Chap. 5)

    Stat → ANOVA → One-way
    Stat → ANOVA → One-way (Unstacked)

ANOVA, two-way design (Chap. 5)

    Stat → ANOVA → General Linear Model →

       Fit General Linear Model

Bar chart (Chap. 2)

    Graph → Bar Chart

Bartlett's test – See Homogeneity of variance

Box-and-whisker plot (Chap. 2)

    Graph → Boxplot

Chi square tests (Chap. 6)

    Stat → Tables → Chi Square Test (Table in Worksheet)
    Stat → Tables → Cross Tabulation and Chi Square
    Stat → Tables → Chi Square Goodness-of-Fit Test (one variable)

Cochran-Mantel-Haenszel test (Chap. 6)

    Stat → Tables → Cross Tabulation and Chi Square → *Other stats*

Coefficient of variation – See Descriptive statistics

Confidence intervals

    Z-distribution – See Z-test, one sample
    t-distribution – See t-test, one sample

Correlation (Chap. 6)

    Stat → Basic Statistics → Correlation

Descriptive statistics (Chap. 2)

    Stat → Basic Statistics → Display Descriptive Statistics

Dixon's test (Chap. 8)

    Stat → Basic Statistics → Outlier test → *Options*

Dot chart (Chap. 2)

    Graph → Dot plot

Dunnett's test – See post hoc procedures

Fisher exact test (Chap. 6)

    Stat ➜ Tables ➜ Cross Tabulation and Chi Square ➜ *Other stats*

Fisher's LSD test – See post hoc procedures

Friedman test (Chap. 5)

    Stat ➜ Nonparametrics ➜ Friedman

Grubbs' test (Chap. 8)

    Stat ➜ Basic Statistics ➜ Outlier test ➜ *Options*

Histogram (Chap. 2)

    Graph ➜ Histogram

Homogeneity of variance (Chap. 4) – Bartlett's and Levene's tests

    Stat ➜ ANOVA ➜ Test for equal variances

Hsu's MCB test – See post hoc procedures

Kruskal-Wallis test (Chap. 5)

    Stat ➜ Nonparametrics ➜ Kruskal-Wallis

Kurtosis (Chap. 2)

    Stat ➜ Basic Statistics ➜ Display Description Statistics

Levene's test – See Homogeneity of variance

Linear regression (Chap. 6)

    Stat ➜ Regression ➜ Fit Line Plot

Logistic regression (Chap. 6)

    Stat ➜ Regression ➜ Binary Logistic Regression

Mann-Whitney test (Chap. 5)

    Stat ➜ Nonparametrics ➜ Mann-Whitney

McNemar's test (Chap. 6)

    Stat ➜ Tables ➜ Cross Tabulation and Chi Square ➜ *Other stats*

Mean – See Descriptive statistics

Median – See Descriptive statistics

Mode – See Descriptive statistics

Multiple Comparison tests – See post hoc procedures

Multiple regression (Chap. 6)

    Stat ➜ Regression ➜ Regression ➜ Fit Regression Model

Normality tests (Chap. 3)

Anderson-Darling, Kolmogorov-Smirnov and Ryan-Joiner

    Stat ➜ Basic Statistics ➜ Normality Test

Pie chart (Chap. 2).

    Graph ➜ Pie Chart

Post hoc procedures (Chap. 5)

Tukey's HSD, Fisher's LSD, Dunnett's and Hsu's MCB tests

    Stat ➜ ANOVA ➜ One-way ➜ *Comparisons...*

Power determination (Chaps. 4 and 5)

    Stat ➜ Power and Sample Size

Random sampling (Chap. 1)

    Calc ➜ Random Data ➜ Sample from Columns

Range – See Descriptive statistics

Sample size determination (Chaps. 4 and 5)

    Stat ➜ Power and Sample Size

Scatter plot (Chap. 2)

    Graph ➜ Scatterplot

Sign test, one-sample (Chap. 5)

    Stat ➜ Nonparametrics ➜ 1-Sample Sign

Skew (Chap. 6)

    Stat ➜ Basic Statistics ➜ Display Description Statistics

Spearman rho test (Chap. 6)

    Stat ➜ Tables ➜ Cross tabulation and Chi Square ➜ *Other Stats...*

Standard deviation – See descriptive statistics

Stem-and-leaf plot (Chap. 2)

    Graph ➜ Stem-and-Leaf

Tolerance Limits (Chap. 3)

    Stat ➜ Quality Tools ➜ Tolerance Interval

t-test, one-sample (Chap. 5)

    Stat ➜ Basic Statistics ➜ 1-sample t

t-test, paired data (Chap. 5)

    Stat ➜ Stat ➜ Basic Statistics ➜ Paired t

t-test, two-sample (Chap. 5)

    Stat ➜ Basic Statistics ➜ 2-sample t

Tukey's HSD test – See post hoc procedures

Variance – See Descriptive statistics

Wilcoxon test, one-sample (Chap. 5)

    Stat ➜ Nonparametrics ➜ 1-Sample Wilcoxon

Z-test of proportions, one-sample (Chap. 6)

    Stat ➜ Basic Statistics ➜ 1-proportion

Z-test of proportions, two-sample (Chap. 6)

    Stat ➜ Basic Statistics ➜ 2-proportions

Z-test, one sample (Chap. 3)

    Stat ➜ Basic Statistics ➜ 1-sample Z

Appendix D: Calculations for Statistical Results Present in Various Chapters

The following are worked out mathematical calculations for many of the Minitab results presented in the previous chapters. Slight deviations from between the two results are due to rounding and not carrying the hand calculations to as many significant figures as the compute program.

Chapter 2

Descriptive statistics for initial sample of 120 assays:

$$\bar{X} = \frac{\sum_{i=1}^{n} x_i}{n} = \frac{49.5 + 46.6 + \dots 50.2}{120} = \frac{6044}{120} = 50.367 \, \text{mg}$$

$$S^2 = \frac{\sum_{i=1}^{n}\left(x_i - \bar{X}\right)^2}{n-1} = \frac{\left(49.5 - 50.367\right)^2 + \dots \left(50.2 - 50.367\right)^2}{119}$$

$$S^2 = \frac{\sum_{i=1}^{n}\left(x_i - \bar{X}\right)^2}{n-1} = \frac{882.33}{119} = 7.415$$

$$S = \sqrt{S^2} = \sqrt{7.415} = 2.723 \, \text{mg}$$

$$\%\text{RSD} = \frac{S}{\bar{X}} \times 100\% = \frac{2.723}{50.367} \times 100 = 5.406\%$$

Median is the average of the two center values (59th and 60th in ascending order):

$$\text{Median} = \frac{50.0 + 50.0}{2} = 50.0 \, \text{mg}$$

Descriptive statistics for deliverable volume (for all 30 measurements):

$$\bar{X} = \frac{\sum_{i=1}^{n} x_i}{n} = \frac{236.4 + 237.3 + \dots 240.3}{30} = \frac{7173.3}{30} = 239.11 \, \text{ml}$$

$$S^2 = \frac{\sum_{i=1}^{n}\left(x_i - \bar{X}\right)^2}{n-1} = \frac{\left(236.4 - 239.11\right)^2 + \dots \left(240.3 - 239.11\right)^2}{29}$$

$$S^2 = \frac{\sum_{i=1}^{n}(x_i - \bar{X})^2}{n-1} = \frac{273.33}{29} = 9.43$$

$$S = \sqrt{S^2} = \sqrt{9.43} = 3.07 \, \text{ml}$$

$$\%RSD = \frac{S}{\bar{X}} \times 100\% = \frac{3.07}{239.11} \times 100 = 1.28\%$$

Median is the average of the two center values (15th and 16th in ascending order):

$$\text{Median} = \frac{239.4 + 239.5}{2} = 239.45 \, \text{ml}$$

Chapter 3

Combinations for 20 objects taken 3 at a time:

$$\binom{20}{3} = \frac{20!}{3! \times 17!} = \frac{20 \times 19 \times 18 \times 17!}{3 \times 2 \times 1 \times 17!} = 1,140$$

Initial sample of tables from a scale-up run ($\sum x_i = 3707$):

$$\bar{X} = \frac{\sum_{i=1}^{n} x_i}{n} = \frac{3707}{20} = 185.35 \, \text{mg}.$$

$$S^2 = \frac{\sum_{i=1}^{n}(x_i - \bar{X})^2}{n-1} = \frac{(187 - 185.35)^2 + \ldots (197 - 185.35)^2}{19} = 47.41$$

$$S = \sqrt{S^2} = \sqrt{47.41} = 6.88 \, \text{mg}.$$

Skew:

$$\text{Skew} = \frac{n}{(n-1)(n-2)} \sum \left[\frac{x_i - \bar{X}}{S} \right]^3$$

$$\text{Skew} = \frac{20}{(19)(18)} \times \left[\left(\frac{187 - 185.4}{6.9} \right)^3 + \ldots \left(\frac{197 - 185.4}{6.9} \right)^3 \right]$$

$$\text{Skew} = 0.058 \times 8.729 = 0.51$$

Confidence interval:
(Z-value for 95% CI = 1.96)

$$\mu = \bar{X} \pm Z_{\alpha/2} \times \frac{\sigma}{\sqrt{n}}$$

$$\mu = 185.4 \pm \left(1.96 \times \frac{7.0}{\sqrt{20}}\right) = 185.4 \pm 3.07$$

$$182.22 < \mu < 188.47 \text{mg.}$$

Tolerance limits: – K-value off table = 2.76 (Odeh and Owens 1980)

$$\text{LTL} = \bar{X} - KS = 185.4 - (2.76 \times 6.9) = 166.36$$

$$\text{UTL} = \bar{X} + KS = 185.4 + (2.76 \times 6.9) = 204.99$$

$$166.36 < TI < 204.99 \text{ mg.}$$

Chapter 4

Comparison of precision for two methods:

$$S_{\text{ratio}}^2 = \frac{S_{\text{New method}}^2}{S_{\text{Established Method}}^2} = \frac{3.151}{4.215} = 0.748$$

From Table B3, $F_{.95,11,11} = 2.818$; therefore:

$$F_{0.05,14,9} = \frac{1}{2.818} = 0.355$$

Upper and lower limits of the 90% confidence interval are:

$$\text{Lower Limit}\left(M_\text{L}\right) = \frac{0.748}{2.818} = 0.265$$

$$\text{Lower Limit}\left(M_\text{U}\right) = \frac{0.748}{0.355} = 2.107$$

Propagation of error example:

$$S_{\text{Total}} = \sqrt{S_1^2 + S_2^2 + \ldots S_k^2} = \sqrt{(1.5)^2 + (3.21)^2} = 3.54$$

Chapter 5

One-sample t-test with chewable gels (confidence interval):

$$\mu = \bar{X} \pm t_{(1-\alpha/2)} \times \frac{S}{\sqrt{n}}$$

$$\mu = 102.45 \pm 2.2622 \times \frac{3.86}{\sqrt{10}}$$

$$\mu = 102.45 \pm 2.76$$

$$99.69 < \mu < 105.21$$

Ratio method:

$$t = \frac{\bar{X} - \mu}{\dfrac{S}{\sqrt{n}}} = \frac{102.45 - 100}{\dfrac{3.86}{\sqrt{10}}} = \frac{2.45}{1.22} = 2.01$$

Two-sample t-test for nicotine patches (confidence interval):

$$S_p^2 = \frac{(n_1 - 1)S_1^2 + (n_2 - 1)S_2^2}{n_1 + n_2 - 2}$$

$$S_p^2 = \frac{(19)(325)^2 + (17)(310)^2}{20 + 18 - 2} = \frac{2006875 + 1633700}{36} = 101{,}127$$

$$\mu_1 - \mu_2 = (\bar{X}_1 - \bar{X}_2) \pm t_{n_1 + n_2 - 2}(1 - \alpha/2) \times \sqrt{\frac{S_p^2}{n_1} + \frac{S_p^2}{n_2}}$$

$$\mu_1 - \mu_2 = (2301 - 2561) \pm 2.028 \times \sqrt{\frac{101{,}127}{20} + \frac{101{,}127}{18}}$$

$$\mu_1 - \mu_2 = -260 \pm 2.028 \times 103.32 = -260 \pm 209$$

$$-469 < \mu < -51$$

Ratio method:

$$t = \frac{\bar{X}_1 - \bar{X}_2}{\sqrt{\dfrac{S_p^2}{n_1} + \dfrac{S_p^2}{n_2}}} = \frac{2301 - 2561}{\sqrt{\dfrac{101{,}127}{20} + \dfrac{101{,}127}{18}}} = \frac{-260}{103.32} = -2.52$$

Paired t-test for contents of a raw material:

$$S_d^2 = \frac{n(\Sigma d^2) - (\Sigma d)^2}{n(n-1)} = \frac{10(2.31) - (-1.9)^2}{10(9)} = \frac{19.49}{90} = 0.217$$

$$S_d = \sqrt{S_d^2} = \sqrt{0.217} = 0.466$$

$$\bar{X}_d = \frac{\Sigma d}{n} = \frac{1.9}{10} = 0.19$$

$$\mu_d = \bar{X}_d \pm t_{n-1}(1-\alpha/2) \times \frac{S_d}{\sqrt{n}}$$

$$\mu_d = 0.19 \pm 2.2622 \times \frac{0.466}{\sqrt{10}} = 0.19 \pm 0.333$$

$$-0.143 < \mu < +523$$

Ratio method:

$$t = \frac{\bar{X}_d}{\dfrac{S_d}{\sqrt{n}}} = \frac{0.19}{0.147} = 1.29$$

One-way ANOVA for dissolution apparatuses:

Tester	N	Mean	StDev	95% CI
A	18	80.578	2.610	(79.341, 81.814)
B	18	77.578	2.469	(76.341, 78.814)
C	24	78.221	2.478	(77.150, 79.292)
D	24	77.387	2.764	(76.317, 78.458)
E	18	75.439	2.881	(74.203, 76.675)

$$MS_w = \frac{(n_1 - 1)S_1^2 + (n_2 - 1)S_2^2 + \ldots (n_k - 1)S_3^2}{N - K}$$

$$MS_w = \frac{(17)(2.610)^2 + (17)(2.469)^2 + \ldots (17)(2.881)^2}{102 - 5} = 6.984$$

$$\bar{X}_G = \frac{(n_1 \bar{X}_1) + (n_2 \bar{X}_2) + \ldots (n_k \bar{X}_k)}{N}$$

$$\bar{X}_G = \frac{(18)(80.578)+(18)(77.578)+...(18)(75.439)}{102} = 77.836$$

$$MS_B = \frac{n_1(\bar{X}_1-\bar{X}_G)^2 + n_2(\bar{X}_2-\bar{X}_G)^2 +...n_k(\bar{X}_k-\bar{X}_G)^2}{K-1}$$

$$MS_B = \frac{(18)(80.578-77.836)^2 +...(18)(75.439-77.836)^2}{4} = 62.087$$

$$F = \frac{62.087}{6.984} = 8.89$$

Multiple comparisons for dissolution apparatuses:
Tukey (Tester C vs. A): $q = 2.78$ (Pearson and Hartley 1970)

$$\mu_C - \mu_A = (\bar{X}_C - \bar{X}_A) \pm q_{\alpha,k,N-k}\sqrt{\frac{MS_W}{n_C}+\frac{MS_W}{n_C}}$$

$$\mu_A - \mu_C = (78.221-80.578) \pm 2.78\sqrt{\frac{6.985}{18}+\frac{6.985}{24}}$$

$$\mu_A - \mu_C = -2.357 \pm 2.291$$

$$0.066 < \mu_A - \mu_C < 4.648$$

Fisher (Tester A vs. C):

$$\mu_A - \mu_C = (\bar{X}_A - \bar{X}_C) \pm t_{1-\alpha/2,N-k}\sqrt{\frac{MS_W}{n_C}+\frac{MS_W}{n_C}}$$

$$\mu_A - \mu_C = (80.578-78.221) \pm 1.984\sqrt{\frac{6.985}{18}+\frac{6.985}{24}}$$

$$\mu_1 - \mu_2 = 2.357 \pm 1.984(0.824) = 2.357 \pm 1.635$$

$$0.722 < \mu_A - \mu_C < 3.992$$

Dunnett (Tester C vs. A): $q = 2.50$ (Dunnett 1955)

$$\mu_{Control} - \mu_A = (\bar{X}_{Control} - \bar{X}_A) \pm q_{\alpha,k,N-k}\sqrt{MS_W 1\left(\frac{1}{n_{Control}}+\frac{1}{n_A}\right)}$$

$$\mu_{Control} - \mu_i = (78.221 - 80.578) \pm 2.50 \sqrt{6.985 \left(\frac{1}{24} + \frac{1}{18} \right)}$$

$$\mu_1 - \mu_2 = -2.357 \pm 2.50(0.824) = -2.357 \pm 2.060$$

$$-4.417 < \mu_A - \mu_{Control} < -0.297$$

One-sample Z-test of proportions (normal approximation):

$$P_O = \hat{p} \pm Z_{(1-\alpha/2)} \sqrt{\frac{\hat{p}(1-\hat{p})}{n}} = 0.05 \pm 1.96 \sqrt{\frac{(0.05)(0.95)}{100}}$$

$$P_0 = 0.05 \pm 1.96(0.0218) = 0.05 \pm 0.043$$

$$0.007 < P_0 < 0.093$$

$$z = \frac{\hat{p} - P_0}{\sqrt{\dfrac{P_0(1-P_0)}{n}}} = \frac{0.05 - 0.012}{\sqrt{\dfrac{(0.012)(0.988)}{100}}} = \frac{0.038}{0.0109} = 3.48$$

Two-sample Z-test of proportions (normal approximation):
High speed $\hat{p}_1 = 17/480 = 0.0354$
Slow speed $\hat{p}_2 = 11/500 = 0.0220$

$$P_1 - P_2 = (\hat{p}_1 - \hat{p}_2) \pm Z_{(\alpha/2)} \sqrt{\frac{\hat{p}_1(1-\hat{p}_1)}{n_1} + \frac{\hat{p}_2(1-\hat{p}_2)}{n_2}}$$

$$P_1 - P_2 = (0.0354 - 0.0220) \pm 1.96 \sqrt{\frac{0.0354(0.9646)}{480} + \frac{0.0220(0.9780)}{500}}$$

$$P_1 - P_2 = (0.0134) \pm 1.96(0.0107) = (0.0134) \pm 0.0210$$

$$-0.0076 < P_1 - P_2 < 0.0344$$

$$z = \frac{\hat{p}_1 - \hat{p}_2}{\sqrt{\dfrac{\hat{p}_1(1-\hat{p}_1)}{n_1} + \dfrac{\hat{p}_2(1-\hat{p}_2)}{n_2}}}$$

$$z = \frac{0.0354 - 0.0220}{\sqrt{\dfrac{0.0354(0.9646)}{480} + \dfrac{0.0220(0.9780)}{500}}} = \frac{0.0134}{0.0107} = 1.25$$

Chapter 6

Correlation for comparing two analytical methods:

	Method A (x)	Method B (y)	x^2	y^2	xy
	55	56	3025	3136	3080
	66	67	4356	4489	4422
	46	45	2116	2025	2070
	77	75	5929	5625	5775
	57	57	3249	3249	3249
	59	59	3481	3481	3481
	70	69	4900	4761	4830
	57	59	3249	3481	3363
	52	51	2704	2601	2652
	36	38	1296	1444	1368
	44	45	1936	2025	1980
	55	56	3025	3136	3080
	53	51	2809	2601	2703
	67	68	4489	4624	4556
	72	74	5184	5476	5328
$\Sigma =$	866	870	51,748	52,154	51,937

$$r = \frac{n \sum xy - \sum x \sum y}{\sqrt{n \sum x^2 - (\sum x)^2} \sqrt{n \sum y^2 - (\sum y)^2}}$$

$$r = \frac{15(51,937) - (866)(870)}{\sqrt{15(51,748) - (866)^2} \sqrt{15(52,154) - (870)^2}}$$

$$r = \frac{779,055 - 753,420}{(162.06)(159.41)} = \frac{25,635}{25,833.98} = +.992$$

Linear regression (stability example):

	Time (x)	Assay (y)	x^2	y^2	xy
	6	995	36	990,025	5970
	12	984	144	968,256	11,808
	18	973	324	946,729	17,514
	24	960	576	921,600	23,040
	36	952	1296	906,304	34,272
	48	948	2304	898,704	45,504
Sum =	144	5812	4680	5,631,618	138,108

$$b = \frac{n\sum xy - (\sum x)(\sum y)}{n\sum x^2 - (\sum x)^2} = \frac{(6)(138,108) - (144)(5812)}{(6)(4680) - (144)^2}$$

$$b = \frac{-8280}{7344} = -1.127$$

$$a = \frac{\sum y - b\sum x}{n} = \frac{5812 - (-1.127)(144)}{6} = 995.715$$

Regression line:

$$y = a + bx = 995.715 - 1.127x$$

Coefficient of determination:

$$SS_T = \sum y^2 - \frac{(\sum y)^2}{n} = 5,631,618 - \frac{(5812)^2}{6} = 1727.33$$

$$SS_E = b^2\left[\sum x^2 - \frac{(\sum x)^2}{n}\right] = (-1.127)^2 \times \left[4680 - \frac{(144)^2}{6}\right]$$

$$SS_E = 1554.64$$

$$SS_U = SS_T - SS_E = 1727.33 - 1554.64 = 172.69$$

$$r^2 = \frac{SS_E}{SS_T} = \frac{1554.64}{1727.33} = 0.900$$

ANOVA table:

Source	df	SS	MS	F	p*
Line	1	1554.64	1554.64	36.01	0.004
Error	4	172.69	43.17		
Total	5	1727.33			

*p-value calculated using Excel function F.DIST.RT = 36.01

Confidence interval at 48 months:

Point on the regression line at 48 months:

$$y = a + bx = 995.715 + (-1.127)(48) = 941.619$$

$$\bar{y} = y_c \pm t_{n-2,1-\alpha/2} \times \sqrt{MS_{residual}} \sqrt{\frac{1}{n} + \frac{\left(x_i - \bar{X}\right)^2}{\sum x^2 - \frac{\left(\sum x\right)^2}{n}}}$$

$$\bar{y} = 941.619 \pm 2.7765 \times \sqrt{43.17} \sqrt{\frac{1}{6} + \frac{\left(48-24\right)^2}{4680 - \frac{\left(144\right)^2}{6}}}$$

$$\bar{y} = 941.619 \pm 14.557$$

$$927.062 < \bar{y} < 956.176$$

Chi square:
Initial analysis (5 × 3 table) – Expected values

1	5.67	37.3	5	1
1	5.67	37.3	5	1
1	5.67	37.3	5	1

$$\chi^2 = \sum \frac{\left(O-E\right)^2}{E} = \frac{\left(0-1\right)^2}{1} + \frac{\left(8-5.67\right)^2}{5.67} + \ldots \frac{\left(1-1\right)^2}{1} = 14.93$$

Second analysis (3 × 3 table) – expected values:

6.67	37.3	6
6.67	37.3	6
6.67	37.3	6

$$\chi^2 = \sum \frac{\left(O-E\right)^2}{E} = \frac{\left(8-6.67\right)^2}{6.67} + \frac{\left(35-37.3\right)^2}{37.3} + \ldots \frac{\left(3-6\right)^2}{6} = 10.02$$

Fisher's exact test:
For 2 in cell a:

$$p = \frac{\left(a+b\right)! \times \left(c+d\right)! \times \left(a+c\right)! \times \left(b+d\right)!}{N! \times a! \times b! \times c! \times d!}$$

$$p_2 = \frac{7! \times 13! \times 10! \times 10!}{21! \times 2! \times 5! \times 8! \times 6!} = 0.146$$

For 1 in cell a:

$$p_1 = \frac{7! \times 13! \times 10! \times 10!}{21! \times 1! \times 6! \times 9! \times 4!} = 0.027$$

For 0 in cell a:

$$p_0 = \frac{7! \times 13! \times 10! \times 10!}{21! \times 0! \times 7! \times 10! \times 3!} = 0.0015$$

Fisher's exact test (two tailed):

$$p = 2(p_2 + p_1 + p_0) = 2(0.146 + 0.027 + 0.0015) = 0.349$$

McNemar test:

$$\hat{\delta} = \frac{b-c}{n} = \frac{13-2}{30} = 0.367$$

$$\delta = \hat{\delta} \pm \left(Z_{\alpha/2} \times \frac{\sqrt{b+c-n\hat{\delta}^2}}{n} + \frac{1}{n} \right)$$

$$\delta = 0.367 \pm \left(1.96 \times \frac{\sqrt{13+2-30(0.367)^2}}{30} + \frac{1}{30} \right)$$

$$\delta = 0.367 \pm 0.255$$

$$0.112 < \delta < 0.622$$

Cochran-Mantel-Haenszel test:

$$\chi^2_{\text{CMH}} = \frac{\left[\sum \dfrac{a_i d_i - b_i c_i}{N_i} \right]^2}{\sum \left[\dfrac{(a_i+b_i)(c_i+d_i)(a_i+c_i)(b_i+d_i)}{N_1^2(n_1-1)} \right]}$$

$$\chi^2_{\text{CMH}} = \frac{\left[\dfrac{(57)(4)-(3)(56)}{120}+\dfrac{(58)(6)-(2)(54)}{120}\right]^2}{\dfrac{(60)(60)(113)(7)}{(120)^2(119)}+\dfrac{(60)(60)(112)(8)}{(120)^2(119)}} = 1.76$$

Chapter 7

One-sample equivalence test:

$D = 99.85\text{--}100\ (\text{target}) = -0.15;\ df = 5\ (6\text{--}1);$

$SE = S/\sqrt{n} = 0.347,\ t_5(0.95) = 2.02$

Confidence intervals:

$$D_{\text{L}} = D - t_{1-\alpha,\upsilon} \times SE$$

$$D_{\text{L}} = (-0.15) - (2.02)(0.347) = (-0.15) - 0.701 = -0.85$$

$$D_{\text{U}} = D + t_{1-\alpha,\upsilon} \times SE$$

$$D_{\text{U}} = (-0.15) + (2.02)(0.347) = (-0.15) + 0.701 = +0.55$$

Ratio t-statistic:

$$t_1 = \frac{D - \delta_1}{SE} = \frac{(-0.15)-1.5}{0.347} = \frac{-1.65}{0.347} = -4.75$$

$$t_2 = \frac{D - \delta_2}{SE} = \frac{(-0.15)+1.5}{0.347} = \frac{1.35}{0.347} = +3.89$$

Two-sample equivalence test:

$D = \tilde{X}_T - \tilde{X}_R = -1.050;\ df = 10\ (6+6-2);\ 9\ df\ \text{used because of unequal variances}$

$SE = \sqrt{S_p^2/n} = 0.782,\ t_9(0.95) = 1.833$

Confidence intervals:

$$D_{\text{L}} = D - t_{1-\alpha,\upsilon} \times SE$$

$$D_{\text{L}} = (-1.050) - (1.833)(0.782) = (-1.050) - (1.433) = -2.483$$

$$D_{\text{U}} = D + t_{1-\alpha,\upsilon} \times SE$$

$$D_U = (-1.050) + (1.833)(0.782) = (-1.050) + (1.433) = 0.383$$

Ratio t-statistic:

$$t_1 = \frac{D - \delta_1}{SE} = \frac{(-1.050) - (-2)}{0.782} = \frac{3.050}{0.782} = 3.900$$

$$t_2 = \frac{D - \delta_2}{SE} = \frac{(-1.050) - (2)}{0.782} = \frac{-0.950}{0.782} = -1.215$$

Paired equivalence test:
$D = \tilde{X}_d = -0.240$; $df = 9$ (10–2);
$SE = S_d / \sqrt{n} = 0.17333$, $t_5(0.95) = 1.8331$
Confidence intervals:

$$D_L = D - t_{1-\alpha,\upsilon} \times SE$$

$$D_L = (-0.240) - (1.8331)(0.17333) = (-0.240) - (0.317) = -0.557$$

$$D_U = D + t_{1-\alpha,\upsilon} \times SE$$

$$D_U = (-0.240) + (1.8331)(0.17333) = (-0.240) + (0.317) = +0.077$$

Ratio t-statistic:

$$t_1 = \frac{D - \delta_1}{SE} = \frac{(-0.240) - (-2)}{0.17333} = \frac{-2.240}{0.17333} = -12.923$$

$$t_2 = \frac{D - \delta_2}{SE} = \frac{(-0.240) - (2)}{0.17333} = \frac{1.760}{0.1733} = 10.155$$

Chapter 8

Example of Grubbs' test for an outlier:
  Sample mean = 99.757, sample standard deviation = 0.705, potential outlier = 97.98

$$T = \frac{\bar{X} - x_1}{S} = \frac{99.757 - 97.98}{0.705} = \frac{1.777}{0.705} = 2.52$$

Example of Dixon's test for an outlier:
  $n = 12$, use r_{21} where $x_1 = 97.98$, $x_3 = 99.35$, $x_{n-1} = 100.26$

$$\frac{x_3 - x_1}{x_{n-1} - x_1} = \frac{99.35 - 97.98}{100.26 - 97.98} = \frac{1.37}{2.28} = 0.601$$

Calculation for studentized residual for potential outlier with regression

MS_E from ANOVA table = 13.758, critical t-value = 2.306, potential outlier = 4.5, 113.7; 4.5 on the regression line: $(y = a + bx) = 106.03$

$$t = \frac{y_i - y_c}{\sqrt{MS_E}} = \frac{113.7 - 106.03}{\sqrt{13.758}} = \frac{7.67}{3.71} = 2.07$$

Table References

Dunnett CW (1955) A multiple comparison procedure for comparing several treatments with a control. J Am Stat Assoc 50(2):1096–1121

Odeh RE, Owen DB (1980) Tables for normal tolerance limits, sampling plans, and screening, Marcel Dekker, Inc., New York, pp. 90–93 and 98–105

Pearson ES, Hartley HO (1970) Biometrika tables for statisticians, Vol. 1 (Table 29). Biometrika Trustees at the University Press, Cambridge

Index

© American Association of Pharmaceutical Scientists 2019
J. E. De Muth, *Practical Statistics for Pharmaceutical Analysis*, AAPS
Advances in the Pharmaceutical Sciences Series 40,
https://doi.org/10.1007/978-3-030-33989-0

Printed in the United States
By Bookmasters